ADVANCED WIRED AND WIRELESS NETWORKS

MULTIMEDIA SYSTEMS AND APPLICATIONS SERIES

Consulting Editor

Borko Furht
Florida Atlantic University

Recently Published Titles:

CONTENT-BASED VIDEO RETRIEVAL: ***A Database Perspective*** by Milan Petkovic and Willem Jonker; ISBN: 1-4020-7617-7

MASTERING E-BUSINESS INFRASTRUCTURE, edited by Veljko Milutinović, Frédéric Patricelli; ISBN: 1-4020-7413-1

SHAPE ANALYSIS AND RETRIEVAL OF MULTIMEDIA OBJECTS by Maytham H. Safar and Cyrus Shahabi; ISBN: 1-4020-7252-X

MULTIMEDIA MINING: ***A Highway to Intelligent Multimedia Documents*** edited by Chabane Djeraba; ISBN: 1-4020-7247-3

CONTENT-BASED IMAGE AND VIDEO RETRIEVAL by Oge Marques and Borko Furht; ISBN: 1-4020-7004-7

ELECTRONIC BUSINESS AND EDUCATION: ***Recent Advances in Internet Infrastructures***, edited by Wendy Chin, Frédéric Patricelli, Veljko Milutinović; ISBN: 0-7923-7508-4

INFRASTRUCTURE FOR ELECTRONIC BUSINESS ON THE INTERNET by Veljko Milutinović; ISBN: 0-7923-7384-7

DELIVERING MPEG-4 BASED AUDIO-VISUAL SERVICES by Hari Kalva; ISBN: 0-7923-7255-7

CODING AND MODULATION FOR DIGITAL TELEVISION by Gordon Drury, Garegin Markarian, Keith Pickavance; ISBN: 0-7923-7969-1

CELLULAR AUTOMATA TRANSFORMS: ***Theory and Applications in Multimedia Compression, Encryption, and Modeling***, by Olu Lafe; ISBN: 0-7923-7857-1

COMPUTED SYNCHRONIZATION FOR MULTIMEDIA APPLICATIONS, by Charles B. Owen and Fillia Makedon; ISBN: 0-7923-8565-9

STILL IMAGE COMPRESSION ON PARALLEL COMPUTER ARCHITECTURES by Savitri Bevinakoppa; ISBN: 0-7923-8322-2

INTERACTIVE VIDEO-ON-DEMAND SYSTEMS: ***Resource Management and Scheduling Strategies***, by T. P. Jimmy To and Babak Hamidzadeh; ISBN: 0-7923-8320-6

MULTIMEDIA TECHNOLOGIES AND APPLICATIONS FOR THE 21st CENTURY: ***Visions of World Experts***, by Borko Furht; ISBN: 0-7923-8074-6

ADVANCED WIRED AND WIRELESS NETWORKS

edited by

Tadeusz A. Wysocki
University of Wollongong, Australia
Arek Dadej
University of South Australia
Beata J. Wysocki
University of Wollongong, Australia

Tadeusz A. Wysocki
School of Electrical, Computer, &
Telecommunications Engineering
University of Wollongong
2522 Wollongong NSW
Australia
email: wysocki@uow.edu.au

Arek Dadej
Institute for Telecommunications Research
University of South Australia
SPRI Building,Mawson Lakes Blvd.
5095 Mawson Lakes SA
Australia
email: arek.dadej@unisa.edu.au

Beata J. Wysocki
School of Electrical, Computer, &
Telecommunications Engineering
University of Wollongong
2522 Wollongong NSW
Australia
email: beata@elec.uow.edu.au

SERIES EDITOR:
Borko Furht
Dept. of Computer Science &
Engineering
Florida Atlantic University
777 Glades Road
P.O. Box 3091
Boca Raton, FL 33431, U.S.A.
borko@cse.fau.edu

Advanced Wired and Wireless Networks
edited by Tadeusz A. Wysocki, Arek Dadej and Beata J. Wysocki
MULTIMEDIA SYSTEMS AND APPLICATIONS Volume 26

Library of Congress Cataloging-in-Publication Data

A C.I.P. Catalogue record for this book is available
from the Library of Congress.

ISBN 0-387-22781-4 e-ISBN 0-387-22792-X
Printed on acid-free paper.

Printed in the United States of America.

SPIN 11055303, 11310501

springeronline.com

CONTENTS

PART 3: PERFORMANCE OF ADVANCED NETWORKS AND PROTOCOLS

PREFACE

We live in the era of information revolution triggered by a widespread availability of Internet and Internet based applications, further enhanced by an introduction of wireless data networks and extension of cellular networks beyond traditional mobile telephony through an addition of the mobile Internet access. The Internet has become so useful in all areas of life that we always want more of it. We want ubiquitous access (anywhere, anytime), more speed, better quality, and affordability. This book aims to bring to the reader a sample of recent research efforts representative of advances in the areas of recognized importance for the future Internet, such as ad hoc networking, mobility support and performance improvements in advanced networks and protocols. In the book, we present a selection of invited contributions, some of which have been based on the papers presented at the 2nd Workshop on the Internet, Telecommunications and Signal Processing held in Coolangatta on the Gold Coast, Australia, in December 2003.

The first part of the book is a reflection of efforts directed towards bringing the idea of ad-hoc networking closer to the reality of practical use. Hence its focus is on more advanced scalable routing suitable for large networks, directed flooding useful in information dissemination networks, as well as self-configuration and security issues important in practical deployments. The second part of the book illustrates the efforts towards development of advanced mobility support techniques (beyond traditional "mobile phone net") and Mobile IP technologies. The issues considered here range from prediction based mobility support, through context transfer during Mobile IP handoff, to service provisioning platforms for heterogeneous networks. Finally, the last part of the book, on performance of networks and protocols, illustrates researchers' interest in questions related to protocol enhancements for improved performance with advanced networks, reliable and efficient multicast methods in unreliable networks, and composite scheduling in programmable/active networks where computing resources are of as much importance to network performance as transmission bandwidth.

The editors wish to thank the authors for their dedication and lot of efforts in preparing their contributions, revising and submitting their chapters as well as everyone else who participated in preparation of this book.

Tadeusz A. Wysocki
Arek Dadej
Beata J. Wysocki

PART 1:

ADVANCED ISSUES IN AD-HOC NETWORKING

Chapter 1

HIGHLY SCALABLE ROUTING STRATEGIES: DZTR ROUTING PROTOCOL

Mehran Abolhasan and Tadeusz Wysocki

Telecommunication and IT Research Institute (TITR) University of Wollongong, NSW 2522, Australia

Abstract In this paper we present a simulation study of a hybrid routing protocol we proposed in our previous work [4] [3]. Our hybrid routing strategy is called Dynamic Zone Topology Routing protocol (DZTR). This protocol has been designed to provide scalable routing in a Mobile Ad hoc Networking (MANET) environment. DZTR breaks the network into a number of zones by using a GPS. The topology of each zone is maintained proactively and the route to the nodes in other zones are determined reactively. DZTR proposes a number of different strategies to reduce routing overhead in large networks and reduce the single point of failure during data forwarding. In this paper, we propose a number of improvements for DZTR and investigate its performance using simulations. We compare the performance of DZTR against AODV, LAR1 and LPAR. Our results show that DZTR has fewer routing overheads than the other simulated routing protocols and achieves higher levels of scalability as the size and the density of the network is increased.

Keywords: Ad hoc Networks,Routing, DZTR, Scalability.

1. INTRODUCTION

Mobile Ad hoc Networks (MANETs) are comprised of end user nodes, which are capable of performing routing in a distributed fashion. This means that these networks do not require a central coordinator or a base station to perform and establish routes. These networks are particularly useful in areas where an infrastructure is not available or difficult to implement. Such areas include the highly dynamic battle field environment, which requires a mobile networking solution and in the search-and-rescue operations where a large rescue team may be searching through a remote area such as a jungle or a desert.

Similar to most infrustructured or wired networks such as the Internet, MANETs employ a TCP/IP networking model. However, the need to provide end-to-end communication in a dynamic environment, along with the limited resources such as bandwidth and power, demands a redefinition of the layers used in the TCP/IP. Currently, research is being carried out across all layers of the TCP/IP model, to design an infrastructure, which will provide reliable and efficient end-to-end communication for MANETs. One challenging, yet highly researched area in MANETs is routing. In MANET, an intelligent routing strategy is required to provide reliable end-to-end data transfer between mobile nodes while ensuring that each user receives certain level of QoS. Furthermore, the routing strategy must minimise the amount of bandwidth, power and storage space used at each end user node. Therefore, traditional routing strategies, such as the link-state and distance vector algorithm, which where intended for wired or infrastructured networks will not work well in dynamic networking environment.

To overcome the problems associated with the link-state and distance-vector algorithms a number of routing protocols have been proposed for MANETs. These protocols can be classified into three different groups: Global/Proactive, On-demand/Reactive and Hybrid. In proactive routing protocols such as FSR [8], DSDV[13] and DREAM[6], each node maintains routing information to every other node (or nodes located in a specific part) in the network. The routing information is usually kept in a number of different tables. These tables are periodically updated and/or if the network topology changes. The difference between these protocols exists in the way the routing information is updated, detected and the type of information kept at each routing table. Furthermore, each routing protocol may maintain different number of tables. On-demand routing protocols such as AODV[7], DSR[11] and LAR[12] were designed to reduce the overheads in proactive protocols by maintaining information for active routes only. This means that routes are determined and maintained for nodes that require to send data to a particular destination. Route discovery usually occurs by flooding a route request packets through the network. When a node with a route to the destination (or the destination itself) is reached, a route reply is sent back to the source node using link reversal if the route request has travelled through bi-directional links, or by piggy-backing the route in a route reply packet via flooding. Hybrid routing protocols such as ZHLS[10], ZRP [9] and SLURP[14] are a new generation of protocol, which are both proactive and reactive in nature. These protocols are designed to increase scalability by allowing nodes with close proximity to work together to form some sort of a backbone to reduce the route discovery overheads. This is mostly achieved by proactively maintaining routes to nearby nodes and determining routes to far away nodes using a route discovery strategy. Most hybrid protocols proposed to date are zone-based, which means that the network is partitioned or seen as a number of zones by each node. Others group nodes into trees or clusters. Hybrid routing protocols have the potential to provide higher scalability than pure reactive or proactive protocols. This is because they attempt to minimise the number of rebroadcasting nodes by defining a structure (or some sort of a backbone), which allows the nodes to work together in order to organise how routing is to be performed. By working together the best or the most suitable nodes can be used to perform route discovery. For example, in ZHLS only the nodes which lead to the gateway nodes rebroadcast the route discovery packets. Collaboration between nodes can also help in maintaining routing information much longer. For example, in SLURP, the nodes within each region (or zone) work together to maintain location information about the nodes, which are assigned to that region (i.e. their home region). This may potentially eliminate the need for flooding, since the nodes know exactly where to look for a destination every time. Another novelty of hybrid routing protocols is that they attempt to eliminate single point of failure and creating bottleneck nodes in the network. This is achieved by allowing any number of nodes to perform routing or data forwarding if the preferred path becomes unavailable.

Most hybrid routing protocols proposed to date are zone-based. In zone-based routing protocols, the network is divided into a number zones, which can be overlapping ones, such as in ZRP, or non-overlapping such as in ZHLS. The disadvantage of ZRP is that if the zone radius is too large the protocol can behave like a pure proactive protocol, while for a small zone radius it behaves like a reactive protocol. Furthermore, the zones are overlapping, which means that each node can belong to a number of different zones, which increases redundancy. The disadvantage of a non-overlapping zone-based protocols such as ZHLS is that the zone partitioning is done at the design stage. This means that all nodes must have preprogrammed zone maps, which are identical for all nodes in the network, or they must obtain a copy of the zone map before routing can occur. Static zone maps can be used in environments where the geographical boundaries of the network are known (or can be approximated). Such environments include: shopping malls, universities or large office buildings, where physical boundaries can be determined and partitioned into a number of zones. However, in environments where the geographical boundaries of the network are dynamic (i.e. can change from time to time as nodes may travel to different regions), a static zone map cannot be implemented. Examples of such networks include: the battlefield where the battle scene may constantly move from one region to another or in search-and-rescue operations in remote areas. In these environments, a dynamic zone topology is required.

In our previous study [3], we proposed DZTR, where we introduced two dynamic zone creation algorithms, which use a number different location tracking strategies to determine routes with the least amount of overheads. In this paper, we propose a number of improvements for DZTR and investigate its performance using simulation technique. We also compare the performance of DZTR with AODV, LAR1, and LPAR[5], under a number of different network scenarios and comment on their scalability in large networks. The rest of this paper is organised as follows. Section II briefly describes the DZTR routing protocols. Section III describes the simulation tool and the parameters used in our simulations. Section IV presents a discussion on the results we obtained from the preliminary simulations. and section V presents the concluding remarks for the paper.

2. DYNAMIC ZONE TOPOLOGY ROUTING

DZTR is a zone based routing protocol is designed to provide scalable routing in large networks with high levels of traffic. The advantage of DZTR over some of the other zone-based routing protocols described in the previous section includes:

- Zones are created dynamically rather than using a static zone map such as in ZHLS. This means that a preprogrammed zone map is not required.

- Each zone only belongs to one zone, which means that information redundancy is reduced, while a more collaborative environment is defined.

- Single-point of failure is reduced, since there is no cluster-head or a root-node. All nodes within each zone work together to determine the best routes with the least amount of overheads, and data forwarding between each zone can still occur without a route failure as long as there is one gateway connecting the two zones.

- A number of location tracking strategies is proposed to determine routes with minimum amount of overheads for a number of different scenarios.

The DZTR routing protocol is made up of three parts. These are Zone Creation, Topology Determination and Location Discovery. The following sections describe each part.

2.1 Zone Creation

In DZTR two different zone creation algorithms are proposed. These are referred to as DZTR1 and DZTR2. In DZTR1, all nodes in the network start off by being in single state mode, which means that they are not members of any zone. When two nodes come within each others transmission range and form a bi-directional link, a zone is created if the following conditions are satisfied:

- Neither node has a zone ID which maps within their transmission range.

- At least one of the nodes are not a gateway node of another zone[1].

To create a zone ID, each node records its current location, speed and battery power and exchange it with the other using a Zone-Query packet. The coordinates of the node with the lower speed will be used as the zone centre point, which is used to create and reserve the zone boundary. If the nodes have the same velocity, then the node with the higher battery power will be used as the centre point. The aim here is to select the node which is expected to last the longest in the calculated zone. This means that the calculated zone will be active for a longer time.

When the node which has the higher stability of the two is determined, each node will then calculate the boundary using the centre point and the transmission range of that node. Note that when a node sends a Zone-Query Packet, it also keeps a copy of this packet and waits for the other node to send its Zone-Query Packet. When the neighbours Zone-Query packet is received, it uses the two packet to create the zone. The node will then exchange the calculated zone ID to ensure that they have agreed on the same zone ID. If the zone IDs are different the zone ID of the least mobile node is used based on the mobility information exchanged during the zone ID exchange phase. The zone ID will be a function of the centre point and the zone radius. We have chosen the zone ID to be the concatenation of the zone centre point and the zone radius. [2]

$$Z_{ID} = f(C, R) = C|R \tag{1}$$

Similar to DZTR1, the zones are geographically bounded by a zone radius. However, in DZTR2, the boundary of the zone is chosen in such a way that all nodes are within transmission range [3]. The advantage of this strategy is that there is no partitioning in each zone. Therefore, there is all nodes within each zone are aware of each other. Another advantage is that each node can update its intrazone with just one beacon message, as there is no need for further rebroadcast to reach all different parts of each zone. However, the zones created in DZTR2 are smaller than DZTR1, which means that the number of zones in DZTR2 maybe significantly higher than DZTR1. This can increase the number of interzone migration when mobility is high, which will require each node to become affiliated with different zones more frequently. Hence, processing overhead and intrazone update may be higher than in DZTR1.

---

[1] One of the two nodes have a neighbouring node which is a zone member

[2] C = coordinates of the centre node (x,y,z), R = transmission range and the | means concatenation. Note that if we assume that R for all nodes are equal, then $ZID = C$

2.2 Topology Determination

In DZTR[3], once each nodes determines its zone ID, it will start to build its intrazone and interzone routing tables. The intrazone topology of each dynamic zone is maintained proactively and the topology and/or routes to the nodes in the interzone is determined reactively.

2.2.1 Intrazone routing. The intrazone network topology is maintained proactively. Each node in the network periodically broadcasts its location information to the other nodes in its intrazone. However, we minimise the number of control packets propagated through the intrazone by setting the frequency at which each node broadcasts its location to be proportional to its mobility and displacement. That is, each node broadcasts its location information through its intrazone if it has travelled (displaced) a minimum distance. This distance is called Minimum Intrazone Displacement (MID). To determine their displacement, each node starts by recording its current location at the startup using a GPS device. It will then periodically check its location (if the node is mobile), and compare it with the previously recorded location. If the distance between the current and the previous location is greater than or equal to MID, then the node will broadcast its location information through the intrazone and set its current location as the new previous location.

We call this updating strategy, Minimum Displacement Update (MDU). The advantage of this updating strategy is that updates are sent more frequently if the location of a node has changed significantly. The disadvantage of sending updates based on mobility alone is that if a node travels back and forward in a small region update packets are still disseminated, however, the topology may have not necessarily changed. Therefore, sending an update packet will be wasteful.

Intrazone update packets will also be sent if any of the following conditions occur:

1 New node comes online

2 Node enters a new zone

3 Node travels more than MID within a zone

4 Intrazone-Update Timer (IUT) expires

2.2.2 Interzone topology creation. The nodes that are situated near the boundary of each zone can overhear update or data packets travelling through the nodes in their neighoubouring zones. These nodes may also be in transmission

[3] When we say DZTR, we refer to both DZTR1 and DZTR2

range of other nodes which are members of another zone. These nodes are referred to as gateway nodes. When a gateway node learns about an existence of another zone, it will broadcast the zone ID of the new zone through its intrazone. This packet is called an Interzone-Update packet (IEZ). This packet includes the gateway nodes node ID, zone ID, location, velocity and learnt zone ID. Therefore, since the gateway includes its velocity and location information, other member nodes can update the information stored in their intrazone table about that gateway node. Hence, the gateways can reset their IUT timer each time they send one of these packets

2.2.3 Interzone migration. When nodes migrate from one zone to another they send a control packet to the previously visited zone, thus leaving behind a trail. The trail information includes the node's current zone ID, location and velocity. The nodes which receive this trail information update their routing tables. Therefore, the nodes in previously visited zone can forward the location request or data packets for the migrating zone to its current zone.

2.3 Location Discovery

When a node has data to send to a particular destination, if the location of the destination is known, DZTR will attempt a number of different location tracking strategies to determine a fresh route to the destination. The location tracking strategy chosen for a known destination will depend on its physical location, velocity and the time of the last previous communication. If the location of the destination is not know, DZTR will initiate Limited Zone-hop Search with Multizone Forwarding (LZS-MF) to determine a route while minimising overhead. To initiate different location tracking strategies, DZTR introduces four different routing scenarios:

(i) Destination is in the intrazone or is a temporary member.

(ii) Destinations ZID or location is known, and it is expected to be in its current zone.

(iii) Destinations ZID or location is known, but its velocity and location information suggest that it could currently lie a number of different neighbouring zones.

(iv) The location or the ZID of the destination is unknown.

When a source has data to send to a particular destination it firstly starts by checking if the destination is located in the intrazone or it is a temporary member. If the destination is found in one of these tables (i.e. case (i)), the source can start sending data since the route to the destination has been predetermined proactively.

If the destination is not found in the intrazone, then the source will consult its Destination History Table (DHT, described further in [3] . If an entry is found in the DHT, the source will check if the destination still maps in its current zone (using the destinations location, velocity and expiration time in the DHT), if the mapping suggests that the destination is still in its current zone (i.e. case (ii)), the source node will use its interzone table to forward the data packet towards the next zone, which leads to the destination zone.

In (iii), the destination's velocity indicates that it may not be in its recorded zone. In this case the destination node can lie in any number of zones. To find the current zone ID (or location) of the destination, the source node unicasts a Zone Request packet with destination's previously recorded location information (i.e. ZREQ-L), to the zone in which the destination was last suspected to be in, using its interzone topology table. When the ZREQ-L packet reaches the destination's suspected zone, the gateway node which have received this packet will first check to see if the destination is still in the intrazone (or a temporary member). If the destination was not found and no location trail is available, the gateway node will calculate a region in which the destination could have migrated to. We call this the Destinations Expected Region (DER)[4], and it is calculated using the destination's previously known velocity and location information. When the DER is calculated, the gateway node will create a new packet, which includes the source node ID and zone ID, destination ID, a sequence number and the DER. This packet is called a Localised Zone Request (LZREQ). The gateway node forwards this packet to all the neighbouring zones which map into the DER. Each gateway node in the receiving zones will check their tables for the destination, if the destination is not found, they will forward this packet to their outgoing (neighbouring) zones which map into the DER. Note that each node only forward the same LZREQ (or ZREQ) packet once. However, each zone may be queried more than once from different entry points (i.e. gateways). This way if there is clustering within each zone, the zones can still be effectively searched. If the destination is found, the destination will send a ZREP packet back towards the source.

In (iv), the destination's current zone is not known. In this case to search the network effectively while ensuring that overheads are kept low, we introduce a new zone searching strategy called Limited Zone-hop Search with Multizone Forwarding (LZS-MF). In this strategy the source node generates a ZREQ-N packet (N denotes no location information is available for the destination). This packet includes the source node ID, zone ID, location, sequence number, neighbouring zone list and a Zone-Hop (ZH) number. The zone hop number defines the number of zones which the ZREQ-N packet can visit before it expires. To

[4]Note the size of the DER is calculated in a similar manner to[12]

search for an unknown destination, the source node begins by setting $ZH = 1$, which means that only the neighbouring zones can be searched. Each time the ZREQ-N discovery produces no results, the source node increments the value of ZH to increase the search area, and the search is initiated again. This search strategy continues until ZH = MAX-COVERAGE-AREA. The advantage of our limited zone-hop search is that if one of the nearby zones has a trail to the destination (or hosts the destination), we avoid searching all the zones in the network. Now, to ensure that not all nodes within each zone are involved in the routing, each time a gateway node in each zone receives a ZREQ-N packet, it uses its interzone topology table to forward the ZREQ-N packet to the nodes, which lead to the neighbouring zones. We call this Multizone Forwarding (MF). In this strategy the source node starts by consulting its interzone topology table to determine the list of neighbouring zones. It will then store the list of neighbouring zones, along with the neighbouring nodes which lead to one of these neighbouring zones. These nodes are the only nodes, which can forward the ZREQ-N packet towards the next neighbour leading to a neighbouring zone. When a ZREQ-N packet reaches a new zone, the receiving node (i.e. the gateway), will first check its routing tables to see if it has a location information about the destination. If no location destination is found and it has not seen the packet before, it will consult its interzone table and forward the ZREQ-N packet with a new list of neighbouring zones and forwarding nodes. The process continues until the ZH limit is reached, the packet timer expires or the destination is found. When the destination is found, it will send a ZREP packet back towards the source node, indicating its current zone, location and velocity. In DZTR, a link failure may not necessarily lead to route failure. This is because data packets can still be forwarded to their destination if there exists a node which leads towards the destination. A route failure will occur and returned back to the source if no such node can be found.

3. SIMULATION MODEL

In this section we describe the scenarios and parameters used in our simulation. We also describes the performance metrics used to compare our routing strategy with a number of existing routing strategies.

3.1 Simulation Environment and Scenarios

Our simulations were carried out using the GloMoSim[1] simulation package. GloMoSim is an event driven simulation tool designed to carry out large simulations for mobile ad hoc networks. The simulations were performed for 50, 100, 200, 300, 400 and 500 node networks, migrating in a 1000m x 1000m area. IEEE 802.11 DSSS (Direct Sequence Spread Spectrum) was used with maximum transmission power of 15dbm at a 2Mb/s data rate. In the MAC

layer, IEEE 802.11 was used in DCF mode. The radio capture effects were also taken into account. Two-ray path loss characteristics was considered as the propagation model. The antenna hight was set to 1.5m, the radio receiver threshold was set to -81 dbm and the receiver sensitivity was set to -91 dbm according to the Lucent wavelan card[2]. Random way-point mobility model was used with the node mobility ranging from 0 to 20m/s and pause time was set to 0 seconds for continuous mobility. The simulation was ran for 200s[5] and each simulation was averaged over eight different simulation runs using different seed values.
Constant Bit Rate (CBR) traffic was used to establish communication between nodes. Each CBR packet was contained 64 Bytes and each packet were at 0.25s intervals. The simulation was run for 20 and 50 different client/server pairs[6] and each session begin at different times and was set to last for the duration of the simulation.

3.2 Implementation Decisions

The aim of our simulation study was to compare the route discovery performance of DZTR under different levels of traffic and node density with a number of different routing protocols. In our simulations, we compare DZTR with LPAR[7], AODV and LAR1. We implemented DZTR on the top of AODV using AODV's existing error recovery strategy, sequence numbering and broadcast ID strategies. The DZTR2 cluster strategy was implemented as the zone creation strategy in order to eliminate partitioning within each zone and also to allow topology maintenance messages (such as Intrazone, Interzone, Trail updates) to occur by using beaconing messages only. Therefore, each packet is exchanged between neighbouring nodes. For example, when a node sends a trail update packet, this packet is also used by its current intrazone members to update their intrazone table (i.e. it is seen as an intrazone update). Similarly, the nodes in the neighbouring zones update their interzone table and the closest gateway to the node which sent the trail update then broadcasts this trail update in its intrazone.

To reduce the number of intrazone updates in DZTR2, each time a node initiates a ZREQ-N, it also uses this packet to update its intrazone and resets its IUT. Furthermore, to minimise the number of interzone updates propagating through each zone, only the closest known gateway rebroadcasts a learnt zone ID. Similarly, during the zone creation phase, a zone reply is only sent by the node which is closest to the zone which sent a zone query. To minimise the

[5]We kept the simulation time lower than the previous chapter due to a very high execution time required for the 50 Flow scenario

[6]Note that the terms Client/Server, src/dest and Flows are used interchangeably

[7]With stable routing enabled[5]

Table 1-1. DZTR Simulation Parameters

IUT	**30s**
Location Check Timer	**3s**
Zone Query Intervals	**30s**
MID	**150m**
Maximum Zone Hops	**7**

routing overhead when location information is not available at the source, we modified the LZS-MF strategy so that during the first cycle of route discovery (i.e. first attempt at route discovery), each retransmitting node only select one node to represent each known zone in the interzone table during further rebroadcasts and each packet cannot re-enter the same zone. Furthermore, the chosen nodes must be further away from the source than the current hop. For example, if there are 6 neighbouring zones, then each retransmitting node will choose at most 6 other retransmitting nodes to further rebroadcast the control packets away from the source. If the first cycle fails, then in the second cycle, all nodes in the interzone table are chosen, which are further away from the source than the current hop. Finally, in the third cycle, all nodes in the interzone table are chosen regardless of their position.

Table 1-1 illustrates the simulation parameters used for DZTR.

3.3 Performance Metrics

The performance of each routing protocol is compared using the following performance metrics.

- Packet Delivery Ratio (vs) Number of nodes
- Normalised control overhead (O/H) (vs) Number of nodes
- End-to-End Delay (vs) Number of nodes

PDR is the Ratio of the number of packet sent by the source node to the number of packets received by the destination node. Normalised control overhead (O/H) presents the ratio of the number of routing packets transmitted through the network to the number of data packets received at the destination. for the duration of the simulation. This metric will illustrate the levels of the introduced routing overhead in the network. Therefore, Packet Delivery Ratio (vs) Number of nodes represents the percentage of data packets that were successfully delivered as the number of nodes was increased for a chosen value of pause time, and Normalised control overhead (O/H) (vs) Number of nodes shows how many control packet were introduced into the network to successfully transmit each data packet to its destination as the number of nodes is increased for a chosen value of pause time. The last metric is used to investigate the changes in end-to-end delay as the number of nodes is increased. Using these metrics, the level of scalability can be determined by the level of PDR or normalised overhead experienced and the shape of the curves. For example, the protocol which have the highest level of PDR and also maintains the flattest curve, has the highest scalability. For normalised overhead we look for the protocol which has the lowest amount of overhead throughout all different node densities. The last metric is used to investigate the changes in end-to-end delay as the number of nodes is increased.

4. RESULTS

In this section we present the worst case (i.e. zero pause time and constant mobility) scenario results we obtained from our simulation. The results for other levels of mobility can be seen in Appendix A. To investigate the worst case scenario behaviour of each routing protocol, we recorded the PDR, normalised routing overhead and the end-to-end delay introduced into the network. We recorded this behaviour for up to 500 nodes in the network.

4.1 Packet Delivery Ratio Results

Fig. 1-1 and 1-2 illustrate the PDR for the 20 Flows and 50 Flow scenarios. In the 20 Flow scenario all routing protocols achieved over 95% packet delivery across all node density levels. This is because the total number of control packets introduced into the network consumes a small portion of the available bandwidth which still leaves a reasonable level of bandwidth for data transmission. However, in the 50 Flow scenario, DZTR outperform all the other routing strategies through all levels of node density. This becomes more evident as the number of nodes are increased to 500 nodes, where the gap between the curve for DZTR and the curve for the other routing protocols becomes wider. It can be seen that at the 500 node density level, AODV, LAR1 and LPAR achieve less than 50% PDR, whereas DZTR achieves over 80%. This is because in DZTR, the increase in the number of nodes may not increase the number of zones in the network. This means that the number of neighbouring zones for each zone may not increase significantly. As a results, the number of retransmitting nodes chosen from the interzone table will remain reasonably low. In contrast, in AODV, LAR1 and LPAR, the increase in node density will increase the number of retransmitting nodes. This will reduce the available bandwidth for data transmission and increase channel contention, which will result in further packet losses due to buffer overflows. Furthermore, in DZTR, a link failure may not initiate a re-discovery of another route, if another gateway node can successfully transmit the data packets. Whereas in AODV, LAR1 and LPAR a link failure may require an alternate route to be discovered. LAR1, attempts to

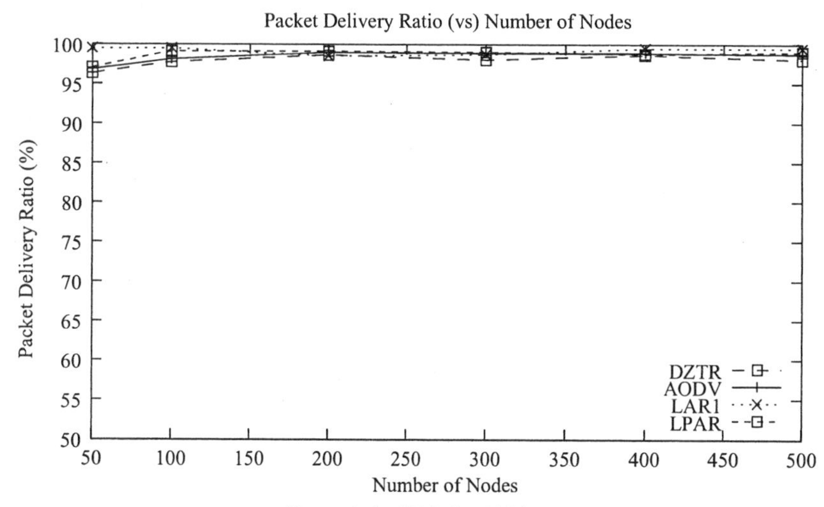

Figure 1-1. PDR for 20 Flows

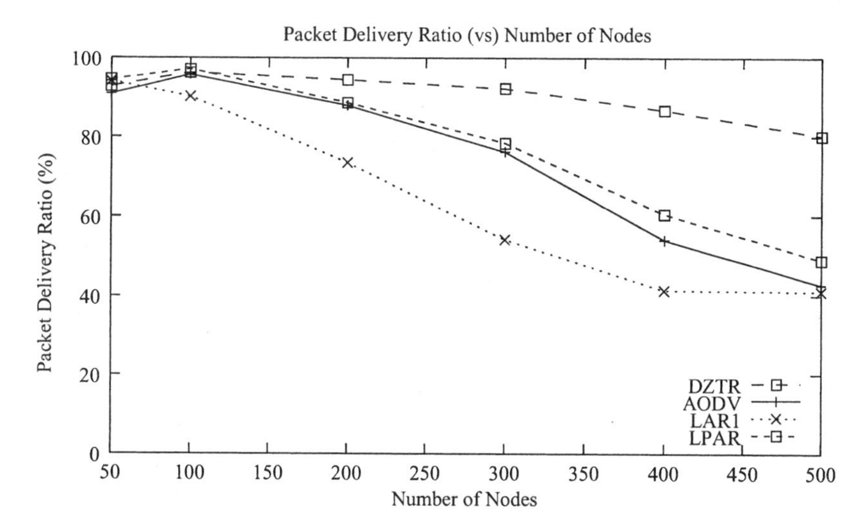

Figure 1-2. PDR for 50 Flows

reduce the number of route recalculations by storing multiple route in a route cache (DSR based). However, since the best route is always used first, then storing alternate route may not be beneficial when mobility is high. Since this route may already be expired or broken when it is required. Hence, in this case, recalculation on alternate route may not be avoided by storing multiple routes. Similarly, in LPAR, the secondary route may expire before a link breaks in the primary route. This means that the alternate route in LPAR may not be always available or valid, especially during high levels of mobility. Therefore, the source nodes may be required to make frequent route recalculations, which will increase the level of bandwidth consumed by routing packets throughout the network.

4.2 Normalised Routing Overhead Results

Fig. 1-3 and 1-4 demonstrate the normalised control overhead for the 20 Flows and the 50 Flows scenarios. In both scenarios DZTR produces the least amount of overhead per packet. Note that as the node density is increased, DZTR maintains the flattest curve when compared to the other three routing strategies, which shows that number of retransmitting nodes do not significantly increase in DZTR. Therefore, the total number of control packets disseminated into the network remains reasonably low as the node density is increased. This shows that DZTR scales significantly better than the other strategies. AODV

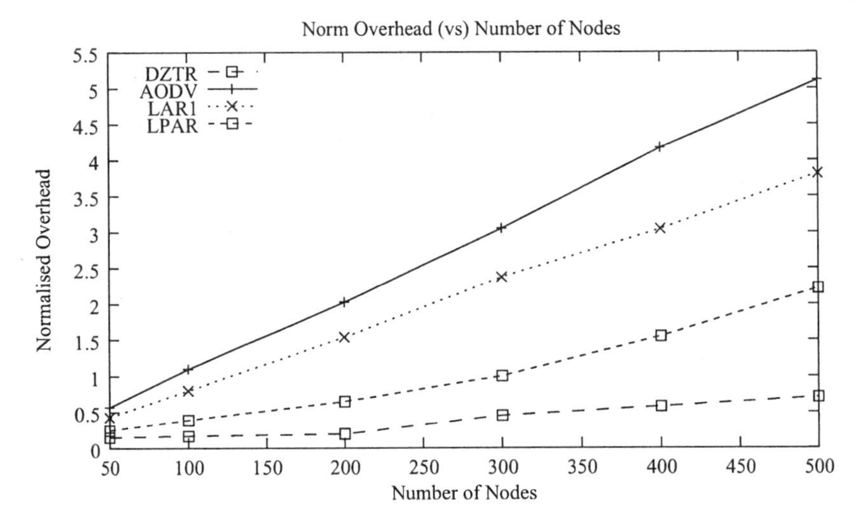

Figure 1-3. Normalised overhead for 20 Flows

produces more overhead that the other strategies across all different levels of node density in the 20 Flow scenario. However, in the 50 Flow scenario AODV and LAR1 produce similar levels of overhead. This is because LAR1 performs source routing rather than point-to-point routing (described in chapter 2 and chapter 4), which means the rate at which route failures occur will be higher than the point-to-point based routing protocols (i.e. AODV, LPAR and DZTR), since the routes are not adaptable to the changes in network topology. Therefore, link failures in LAR1 will initiate more route recalculations at the source than in the point-to-point routing protocols. LPAR produces fewer routing packet than AODV and LAR1 in both of the 20 Flow and the 50 Flow scenario. This reduc-

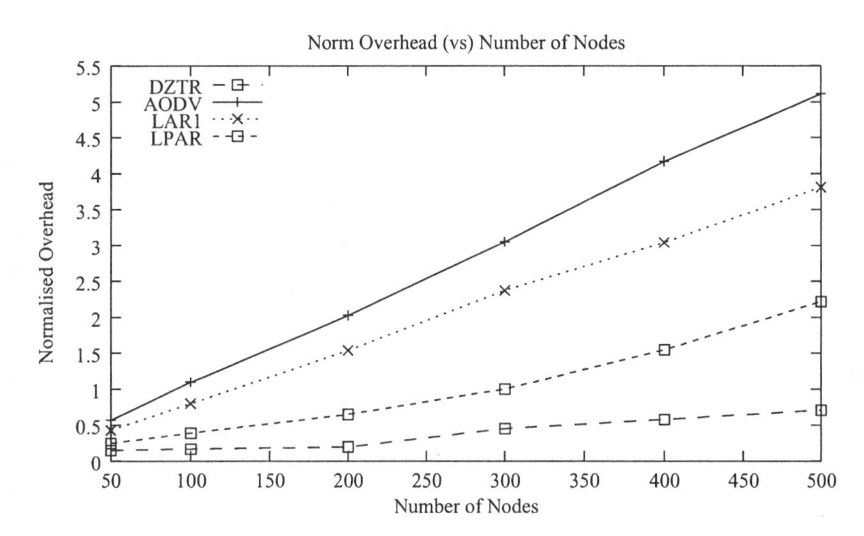

Figure 1-4. Normalised overhead for 50 Flows

tion is achieved by using the 3-state route discovery strategy, which attempts to find a route to a required destination by unicasting if location information about the destination is available (described in chapter 4). Thus reducing the need for broadcasting during route discovery. Furthermore, LPAR reduces the number of control packet retransmission by flooding over stable links only.

4.3 Delay Results

Fig. 1-5 and 1-6 illustrate the end-to-end delay experienced by each data packet for the 20 Flows and the 50 Flows scenarios. In the 20 Flow and 50 Flow scenario for the 50 node network DZTR produces longer delays than the other strategies. Two factor contribute to this extra delay, firstly when the node density is low, the nodes may be engaged in zone creation more frequently as the chance for network partitioning to occur is much higher. This means nodes may go in and out of single-state mode or may become temporary members. Therefore, the information kept in each interzone table may not be very accurate, and the first cycle of route discovery may not always be successful.

The second factor is due to stable routing. In DZTR, a source nodes attempts of find a route over stable links, similar to LPAR, which limits the number of nodes which can rebroadcast. Therefore, more attempts maybe required to determine a route over less stable links. This increase in extra delay can be also seen in LPAR. AODV (which uses Expanding Ring Search, ERS) produces the lowest delay in the 50 node scenario, and maintains similar levels of overhead when compared with DZTR and LPAR. This is because, AODV the flooding nature of AODV allows every node to rebroadcast (if the RREQ packet has not

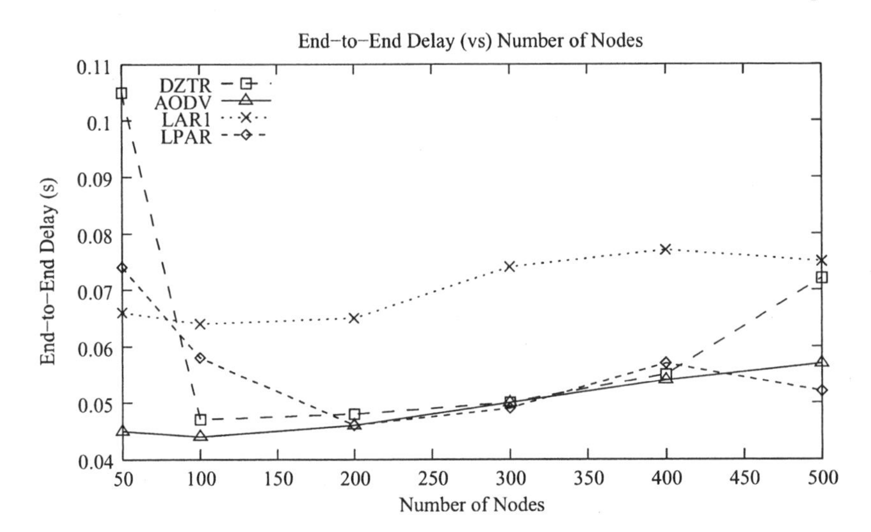

Figure 1-5. End-to-end delay for 20 Flows

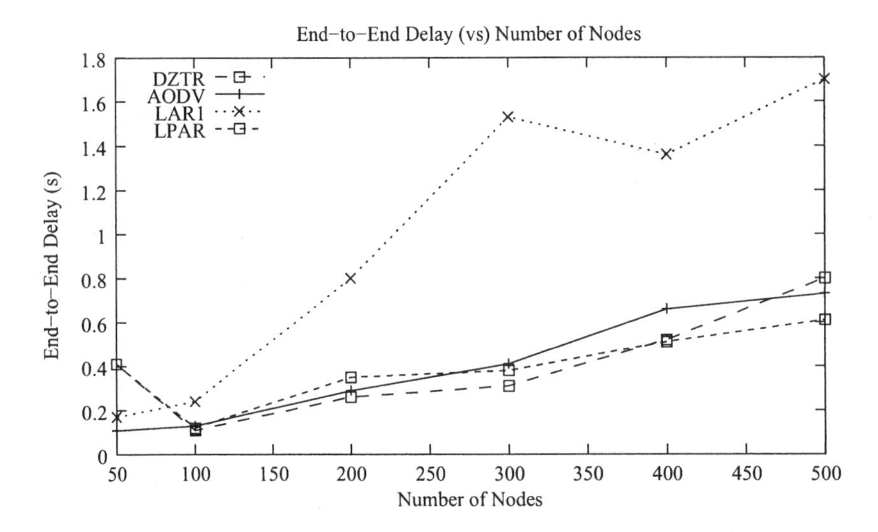

Figure 1-6. End-to-end delay for 50 Flows

expired). Therefore, it calculates the path between the source to the destination more quickly. When the node density is increased to 100, DZTR's end-to-end delay drop dramatically. This is because the higher node density allows DZTR to calculate the required routes more quickly as the LZS-MF becomes more effective in their first route discovery cycles. The delay experienced by all protocols increases slowly as the number of nodes is increased. AODV, LPAR and DZTR experience similar levels of delay for all node density levels greater than 500. However, LAR1 continues to produce larger delays than the other routing protocols during higher node density levels. This is because when mobility is high, more packets may travel over non-optimal routes with larger hop counts, which may be stored in a route cache (described in Chapter 4). Therefore, these packets will experience longer end-to-end delay than the ones travelling over the shortest path. Furthermore, as the node density is increased, the number of routes stored in the route cache may also increase. This means that more non-optimal routes with large hop counts may be available for each required destination. Hence, the probability of longer (non-optimal) end-to-end delay experienced by each packet also increase.

5. CONCLUSIONS

This paper presented a new routing protocol for mobile ad hoc networks, which is called Dynamic Zone Topology Routing (DZTR). The idea behind this protocol is to group nodes that are in close proximity of each other into zones. By grouping nodes together and allowing routing and data transmission to be carried out by a group of nodes, we eliminate single points of failure during data transmission, distribute network traffic through a set of nodes and avoid frequent route recalculation. The topology of each routing zone is maintained proactively and each zone member node is aware of the neighbouring zones through the gateway nodes. DZTR reduces routing overheads by reducing the search zone and allowing only selected nodes to forward the control packets. Each node that migrates between zones also leaves transient zone trails, which assist our proposed search strategy to find the destination more quickly and with fewer overheads. Our theoretical overhead analysis and simulation studies showed that DZTR significantly reduces the number of control packets transmitted into the network and achieves higher levels of packet delivery under worst case network conditions when compared to AODV, LAR1 and LPAR.

REFERENCES

1. Glomosim scalable simulation environment for wireless and wired network systems. In *http://pcl.cs.ucla.edu/projects/glomosim/.*
2. Orinoco pc card. In *http://www.lucent.com/orinoco.*
3. M. Abolhasan, T. Wysocki, and E. Dutkiewicz. Zone-Based Routing Algorithm for Mobile Ad Hoc Networks.
4. M. Abolhasan, T. Wysocki, and E. Dutkiewicz. Scalable Routing Strategy for Dynamic Zone-based MANETs. In *Proceedings of IEEE GLOBECOM*, Taipei, Taiwan, November 17-21 2002.
5. Mehran Abolhasan, Tadeusz Wysocki, and Eryk Dutkiewicz. LPAR: An Adaptive Routing Strategy for MANETs. *In Journal of Telecommunication and Information Technology*, pages 28–37, 2/2003.
6. S. Basagni, I. Chlamtac, V.R. Syrotivk, and B.A. Woodward. A Distance Effect Algorithm for Mobility (DREAM). In *Proceedings of the Fourth Annual ACM/IEEE International Conference on Mobile Computing and Networking (Mobicom'98)*, Dallas, TX, 1998.

7. S. Das, C. Perkins, and E. Royer. Ad Hoc On Demand Distance Vector (AODV) Routing. In *Internet Draft, draft-ietf-manet-aodv-11.txt*, work in progress, 2002.
8. M. Gerla. Fisheye State Routing Protocol (FSR) for Ad Hoc Networks. In *Internet Draft, draft-ietf-manet-aodv-03.txt*, work in progress, 2002.
9. Z.J. Hass and R Pearlman. Zone Routing Protocol for Ad-Hoc Networks. In *Internet Draft, draft-ietf-manet-zrp-02.txt*, work in progress, 1999.
10. Mario Joa-Ng and I-T Lu. A Peer-to-Peer Zone-based Two-level Link State Routing for Mobile Ad Hoc Networks. *IEEE Journal on Selected Areas in Communications*, 17(8), 1999.
11. D. Johnson, D. Maltz, and J. Jetcheva. The Dynamic Source Routing Protocol for Mobile Ad Hoc Networks. In *Internet Draft, draft-ietf-manet-dsr-07.txt*, work in progress, 2002.
12. Yong-Bae Ko and Nitin H. Vaidya. Location-Aided Routing (LAR) in Mobile Ad Hoc Networks. In *Proceedings of the Fourth Annual ACM/IEEE International Conference on Mobile Computing and Networking (Mobicom'98)*, Dallas, TX, 1998.
13. C.E. Perkins and T.J. Watson. Highly Dynamic Destination Sequenced Distance Vector Routing (DSDV) for Mobile Computers. In *ACM SIGCOMM'94 Conference on Communications Architectures*, London, UK, 1994.
14. Seung-Chul Woo and Suresh Singh. Scalable Routing Protocol for Ad Hoc Networks. *accepted for publication in Journal of Wireless Networks (WINET)*, 2001.

Chapter 2

LOCALISED MINIMUM SPANNING TREE FLOODING IN AD-HOC NETWORKS

Justin Lipman[1,2], Paul Boustead[1], Joe Chicharo[1]
[1] *Telecommunications and IT Research Institute, University of Wollongong, Australia,*
[2] *Cooperative Research Centre for Smart Internet Technology, Australia*

Abstract: Information dissemination (flooding) forms an integral part of routing protocols, network management, service discovery and information collection. Given the broadcast nature of ad hoc network communications, information dissemination provides a challenging problem. In this chapter we compare the performance of existing distributed ad hoc network flooding algorithms indentifying stengths and weaknesses inherent in each mechanism. Additionally we propose to apply the Minimum Spanning Tree (MST) algorithm in a distributed manner as the basis of an optimised ad hoc network flooding algorithm called Localised Minimum Spanning Tree Flooding (LMSTFlood). LMSTFlood provides significant reduction in duplicate packet reception, average transmission distance and energy consumed. Thus LMSTFlood limits the broadcast storm problem more effectively than existing optimised flooding mechanisms.

Key words: Flooding, Broadcasting, MST, Localised, Distributed, Ad hoc Network, MANET

1. INTRODUCTION

The advent of portable computers and wireless networking has lead to large growth in mobile computing due to the inherent flexibility offered. Most wireless networks are built around an infrastructure, where all communications is routed through base stations that act as gateways between the wireless and wired network. However, there may be situations in which it is impossible or not desirable to construct such an infrastructure.

An ad hoc network is a collection of wireless mobile nodes forming a temporary network lacking the centralized administration or standard support services regularly available on conventional networks. Nodes in an ad hoc network may act as routers, forwarding packets. Ad hoc networks may undergo frequent changes in their physical topology as mobile nodes may move, thereby changing their network location and link status. New nodes may unexpectedly join the network or existing nodes may unexpectedly leave, move out of range or switch off. Portions of the network may experience partitioning or merging, which is non-deterministic. Ad hoc networks may operate in isolation or connected to a fixed network (Internet) via a base station (gateway). Ad hoc networks are characterised by low bandwidth, high error rates, intermittent connectivity (partitioning), limited transmission range, device power constraints and limited processing capabilities. Most importantly, all communications in an ad hoc network is broadcast in nature, therefore nodes must compete for access to a shared medium.

Information dissemination (flooding) forms an integral part of all communications in ad hoc networks. Given the broadcast nature of ad hoc networks, this poses a challenging problem. It is, therefore, important that any information dissemination mechanism in ad hoc networks be optimised to reduce the problems associated with broadcast communications. In [1], the problems associated with information dissemination in ad hoc networks are identified and refered to as the *broadcast storm problem*. The broadcast storm problem states that flooding is extremely costly and may result in redundant broadcasts, medium contention and packet collisions.

In this chapter, we compare the performance of existing distributed ad hoc network flooding algorithms identifying both strengths and weaknesses of the different approaches. Additionally, we propose to apply the Minimum Spanning Tree (MST) algorithm [2] as the basis of an optimised ad hoc network flooding algorithm called Localised Minimum Spanning Tree Flooding (LMSTFlood). LMSTFlood builds upon work done in [3], where the MST is used for distributed topology control. In LMSTFlood, the MST is determined locally in a distributed manner by each node using local one hop topology information. The use of a distributed MST allows each node to individually determine its closest neighbouring nodes that must be included in any broadcast to ensure continuation of a flood.

This chapter is organised as follows: Section 2 describes published mechanisms for optimised flooding in ad hoc networks. Section 3 explores the distributed MST and proposes the use of distributed MST as the basis of an optimised flooding mechanism. Section 4 describes the simulation environment and provides results and analysis of the proposed optimised flooding mechanism and existing optimised flooding algorithms. Section 5 introduces future work on flooding reliability. Section 6 concludes the chapter.

2. OPTIMISED FLOODING MECHANISMS

In [4] flooding mechanisms which attempt to reduce redundant broadcasts are categorized as probabilistic based, area based and neighbour knowledge based. Probabilistic based approaches require an understanding of network topology to assign rebroadcast probabilities to nodes. Area based approaches assume nodes have a common transmission range, therefore nodes only rebroadcast if they provide sufficient additional coverage. Neighbour knowledge based approaches require that nodes make rebroadcast decisions based upon local neighbourhood knowledge obtained via beacon messages.

The simplest mechanism for information dissemination within a network is *Blind flooding*. Blind flooding is used by routing protocols such as AODV [5] and DSR [6] to perform route discovery. Blind flooding may also be used in network management to distribute state information or in zero start auto-configuration. In Blind flooding, a node broadcasts a packet, which is received by its surrounding neighbours. Each receiving neighbour then verifies that it has not broadcast the packet before. If not, then the packet is rebroadcast. Blind flooding terminates when all nodes have received and rebroadcast the packet. Blind flooding always chooses the shortest path, because it chooses every possible path in parallel. Therefore no other algorithm can produce a shorter delay. Of course this is not quite accurate, as in wireless networks Blind flooding suffers from the broadcast storm problem, which may increase resource contention and hence impede its overall performance.

Multipoint Relay (MPR) flooding, as described in [7], is a distributed two hop neighbour knowledge based flooding mechanism employed in the OLSR routing protocol [8] for the dissemination of link state information. MPR aims to reduce the number of redundant retransmissions during flooding. The number of retransmitters is restricted to a small set of neighbour nodes unlike Blind flooding. This set of nodes is minimized by efficiently selecting one hop neighbours that provide two hop cover of the network area provided by the complete set of one hop neighbours. These selected one hop neighbours are the multipoint relays for a given node. The mechanism is distributed as each node must determine its own MPR set independent of other nodes. Finding the minimal MPR set is NP-complete, however the following algorithm is proposed:

1 Find all 2-hop neighbours reachable from only one 1-hop neighbour. Assign the 1-hop neighbours as MPRs.

2 Determine the resultant cover set - the set of 2-hop neighbours that will receive the packet from the current MPR set.

3 From the remaining 1-hop neighbours not in the MPR set, find the ones that cover the most 2-hop neighbours not in the coverage set.

4 Repeat from step 2 until all 2-hop neighbours are covered.

MPR attempts to minimize the broadcast storm problem by removing redundant broadcasts and grouping nodes into sets which may be reached by relay nodes, thereby greatly reducing the number of rebroadcasting nodes. However, it is also possible to limit the broadcast storm problem by reducing transmission power, thus reducing the broadcast effects and allowing for a reduction in power consumption due to transmission.

Neighbour Aware Adaptive Power (NAAP) flooding [9] is a distributed two hop neighbour knowledge based flooding mechanism for ad hoc networks. NAAP employes several mechanisms (*neighbour coverage*, *transmission power control*, *neighbour awareness* and *local optimisation*) to limit the broadcast storm problem and reduce power consumption in both the transmission and reception of packets during an optimised flood. An intuitive description of the NAAP algorithm is:

1 Upon receiving a broadcast message from a broadcasting node u, each node i (selected by u as a relay) determines which of its one-hop neighbours received the same message.

2 Each relay, i, then determines its closest set of nodes shared with other neighbouring relays and allocates those nodes to its relay set.

3 If nodes in the resulting relay set are not of an equivalent distance from the relay, it may perform a local optimisation on the set to select a minimal subset of relays (with reduced transmission power) that will ensure delivery to remaining nodes in the original optimised set. Otherwise the relay determines a transmission range equal to that of the farthest neighbour it is responsible for.

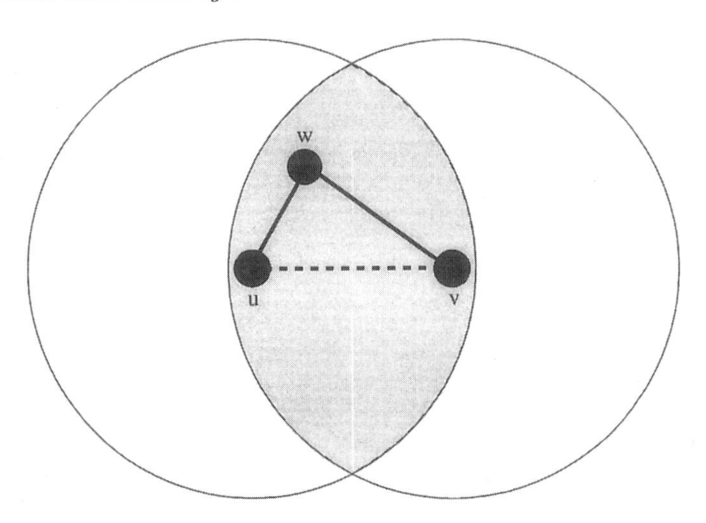

Figure 2-1. Formation of RNG using a Lune

A wireless network may be described by the graph G = (V,E), where V is the set of nodes (vertices) and E the set of edges where $E \subseteq V^2$. Communication between two nodes is possible if an edge (u,v) belongs to E. The distance between two nodes u and v is defined as d(u,v).

The Relative Neighbourhood Graph (RNG) [2] shown in Fig. 2-2 is formed when two nodes are connected with an edge, if their *lune* contains no other nodes of the graph. The lune of two nodes u and v, shown in Fig. 2-1 (in grey) is defined as the intersection of two spheres of radius d(u,v), one centered at node u and the other at node v. The use of a localised RNG was first proposed in [10] as a topology control algorithm to minimize node degrees, hop diameter and maximum transmission range and ensure connectivity. In [11], RNG is applied to flooding in ad hoc networks and is used to address the broadcast storm problem by reducing the transmission range of broadcasting nodes and ensuring the continuation of a flood. Benefits of RNG compared to MPR and NAAP are that the RNG may be determined using local one hop topology information. Nodes in RNG are able to determine whether or not they need to rebroadcast by constructing the RNG. Therefore there is no per packet overhead as with MPR and NAAP.

Optimised flooding mechanisms that utilise transmission power control require a node's location co-ordinates in order to determine the required transmission power. These co-ordinates may be obtained via a positioning system like GPS and shared via periodic exchange of beacon

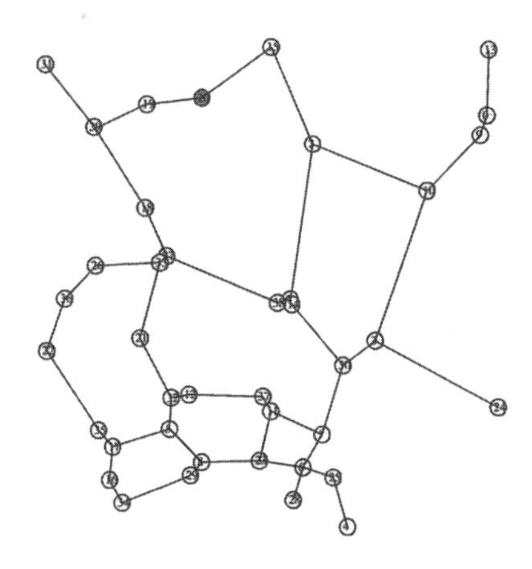

Figure 2-2. Distributed RNG with Local Topology

messages. If a positioning system is not available, distances may be determined through recieved signal strength of beacon messages.

Graphs, such as RNG, in which vertices are connected by an edge, if the edge satisfies some condition of closeness are called proximity graphs. In the next section we propose the use of a popular proximity graph, called the MST, as the basis of a distributed optimised flooding algorithm.

3. LOCALISED MST FLOODING

The Minimum Spanning Tree (MST) graph [2], shown in Fig. 2-3, is a connected graph that uses the minimum total edge length. This results in a graph with one less edge than the number of vertices. The MST is traditionally used in networks for determining broadcast trees using global topology information. The MST is a subgraph of RNG hence the MST may be computed from the RNG by removing edges that create a cycle in the graph. This results in the formation of a tree or directed acyclic graph from all nodes back to the broadcasting node. Thus the MST generates a more optimal broadcast path than RNG, but suffers as there is no fault tolerance in the resulting graph [10]. Fault tolerance refers to the number of alternative paths a message may travel towards a node, thus improving the probability of delivery.

In [3], the authors propose to use the MST algorithm with restricted topology information (one hop) to perform distributed topology control. This is advantageous in ad hoc networks where it is not feasable to

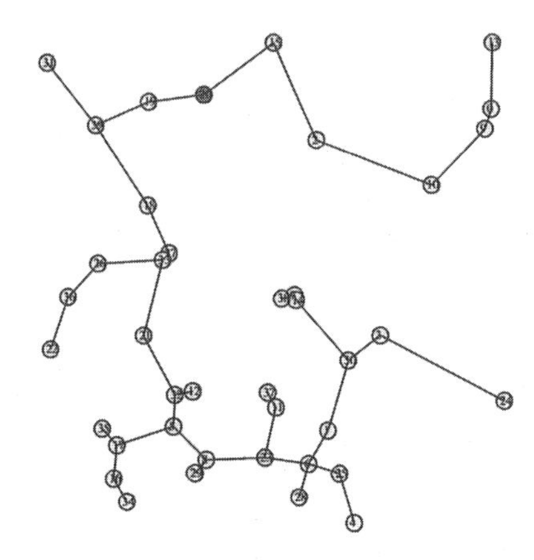

Figure 2-3. Centralised MST with Global Topology

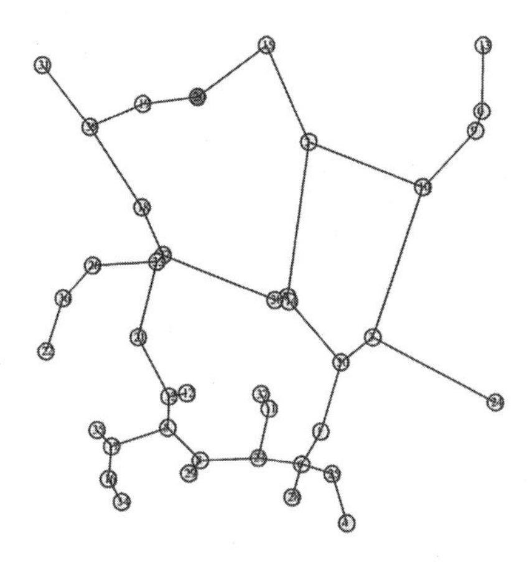

Figure 2-4. Distributed MST with Local Topology

have global topology information for the entire ad hoc network. In this chapter, we propose to apply the MST algorithm in a similar manner to improve the performance of flooding in ad hoc networks. In the distributed MST approach, the topology available to the MST algorithm is restricted to one hop, yet still allows for an optimal broadcast set of nodes with minimal transmission range to be determined as with the centralised approach. Importantly, the resulting distributed MST graph does not exhibit the tree like structure of the centralised MST with global topology knowledge. It can be seen by comparing Figures 2-2, 2-3 and 2-4 that MST $\subseteq$ Localised MST $\subseteq$ RNG as described in [3]. Thus many of the benefits of MST are maintained with the addition of fault tolerance not found in the centralised approach.

Each node, upon receiving a broadcast message, calls algorithm LMSTFlood(). The algorithm determines if the message has been seen before. If not, then a broadcast set (BSET) is determined by supplying the MST with the node's one hop topology information. The previous broadcasting node and all neighbouring nodes that may have heard the previous broadcast are removed from the BSET. If the BSET is not an emptyset, then required transmission power to reach the remaining nodes in the BSET is determined and the message rebroadcast. The MST algorithm used in LMSTFlood() is based upon Prim's algorithm as described in [12].

```
LMSTFlood(Message)
{
  if not seen Message before
  {
    BSET = MST(1-hop Neighbours)
    Last = last node to broadcast
    Heard = determine neighbours that recieved prior broadcast
    BSET = BSET - Last
    BSET = BSET - Heard

    if BSET ≠ ∅
    {
      T_power = maximum_power(BSET)
      Broadcast(Message, T_power)
    }
  }
}
```

4. SIMULATION RESULTS

A simulation was developed that generates a random topology of nodes within a 600 meter by 600 meter area. Nodes have a maximum transmission range of 100 meters. Time is divided into epochs. An ideal MAC layer is assumed. There is no medium contention nor hidden-node scenario within the simulation as it is assumed that during an epoch all nodes can complete their transmission. The transmission medium is error free. A bidirectional link between two nodes is assumed upon reception of a beacon message.

A first order radio model [13] is assumed. In this model the first order radio dissipates $E_{elec} = 50nJ/bit$ to run the circuitry of a transmitter or receiver and a further $E_{amp} = 100pJ/(bit * m^2)$ for the transmitter amplifier. Equation (2.1) is used to calculate the costs of transmitting a k-bit message a distance d. Equation (2.2) is used to calculate the costs of receiving a k-bit message. The radios have power control and consume the minimal required energy to reach the intended recipients.

$$E_{Tx}(k, d) = E_{elec} * k + E_{amp} * k * d^2 \quad (2.1)$$

$$E_{Rx}(k) = E_{elec} * k \quad (2.2)$$

It should be noted that the reasons for using an ideal MAC layer and no mobility are based upon the following: An ideal MAC layer allows us to observe the best case scenario for an optimised flooding mechanism, thus it is possible to determine how effective the mechanism would be at limiting the broadcast storm problem. Additionally, there are various evolving standards for wireless communications other than IEEE 802.11 [14] and therefore an ideal MAC is able to provide us with a generalised idea of performance in a wireless broadcast environment irrespective of the MAC implementation. Mobility is not used as the rate at which a flood progresses throughout the ad hoc network is significantly faster than the change in position of nodes. However, mobility does introduce problems with the accuracy of information available (through beacon messages) to the mechanism when determining whether or not to re-broadcast. Therefore, future work should consider the effects non-ideal MAC layers and mobility have upon optimised flooding mechanisms. More importantly, the reliability of flooding mechanism in the presence of background broadcast and unicast traffic should also be considered.

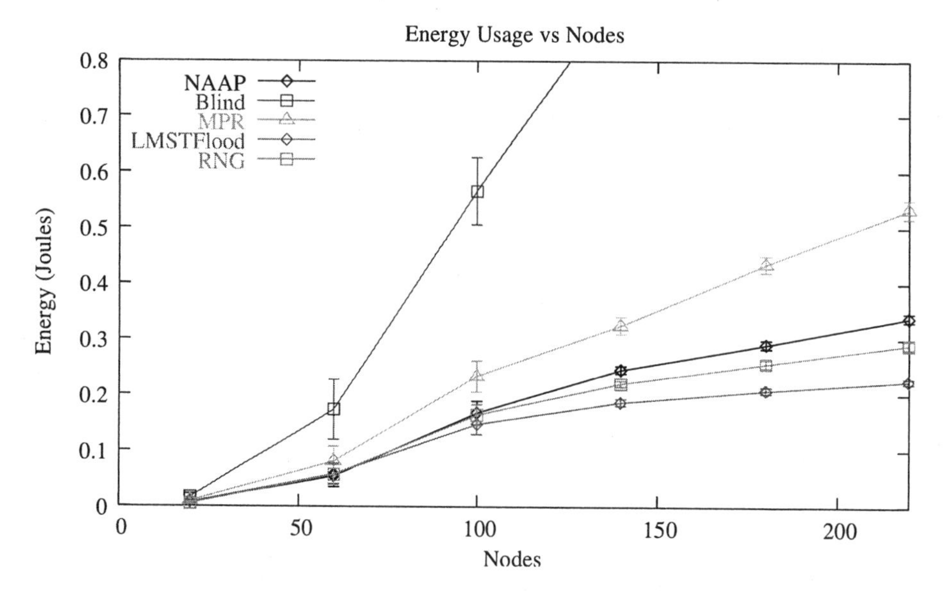

Figure 2-5. A comparison of energy consumed by NAAP, MPR, LMSTFlood, RNG and Blind flooding.

A random node in the topology is selected as the initial node of a flood. Each random topology is used to determine the performance. The topologies generated are not fully connected therefore some topologies may result in a partitioned ad hoc network. The simulation is run 100 times with a different seed for each number of nodes. The results are averaged and 95% confidence intervals are generated. The figures show the performance of each flooding mechanism as the concentration of nodes is increased.

Figure 2-5 shows the energy consumed by each mechanism to complete a flood. The three mechanisms NAAP, RNG and LMSTFlood experience significantly less energy consumption than MPR to complete a flood. From Eq. (2.1), the energy required to transmit a k-bit message is directly proportional to the square of the distance. Therefore, broadcasting over smaller distances is beneficial. Allthough, the number of transmission may be increased as seen in Fig. 2-6, we can see from Eq. (2.2) that there is also a cost associated with recieving a broadcast. In Fig. 2-7 and Fig. 2-8 we see that MPR receives significantly more packets than NAAP, RNG or LMSTFlood. The use of transmission power control when broadcasting allows for a reduction in the number packets recieved by nodes (more importantly the number of duplicate packets

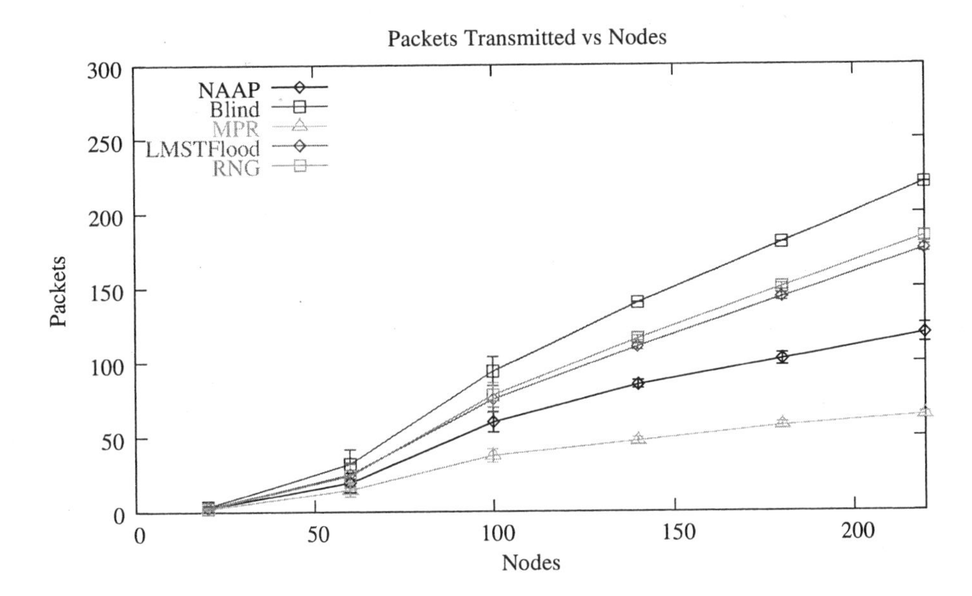

Figure 2-6. A comparison of packets transmitted by NAAP, MPR, LMSTFlood, RNG and Blind flooding.

which are not useful). Only the nearest necessary neighbouring nodes that are required to hear a broadcast will hear it. Thus allowing for a reduction in energy consumption and more effective spatial reuse of the broadcast spectrum.

Figure 2-6 shows the number of transmissions required to complete a flood. All mechanisms show an increase in the number of transmission with respect to the number of nodes. The rate of growth is lower for MPR than for NAAP, RNG and LMSTFlood. This is partially because NAAP, RNG and LMSTFlood attempt to minimize broadcast distance by introducing additional broadcast hops. RNG and LMSTFlood are able to do this more effectively than NAAP as shown in Fig. 2-11. MPR does not control transmission power. NAAP, RNG and LMST-Flood are all able to reduce transmission distance as the node density increases. LMSTFlood shows less transmissions than RNG, this is a result of there being fewer edges assocated with the MST graph compared to the RNG graph (hence less rendundancy) as shown in Fig. 2-2 and Fig. 2-4. As MPR does not use transmission power control, if the density of nodes increases but the network area is maintained then the number of transmissions required to cover all nodes does not grow as

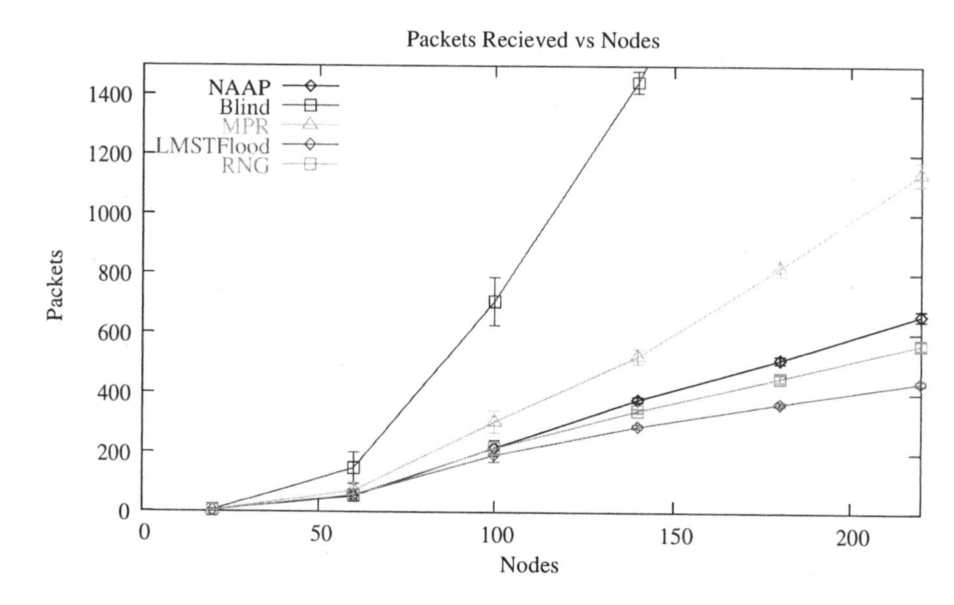

Figure 2-7. A comparison of packets recieved by NAAP, MPR, LMSTFlood, RNG and Blind flooding.

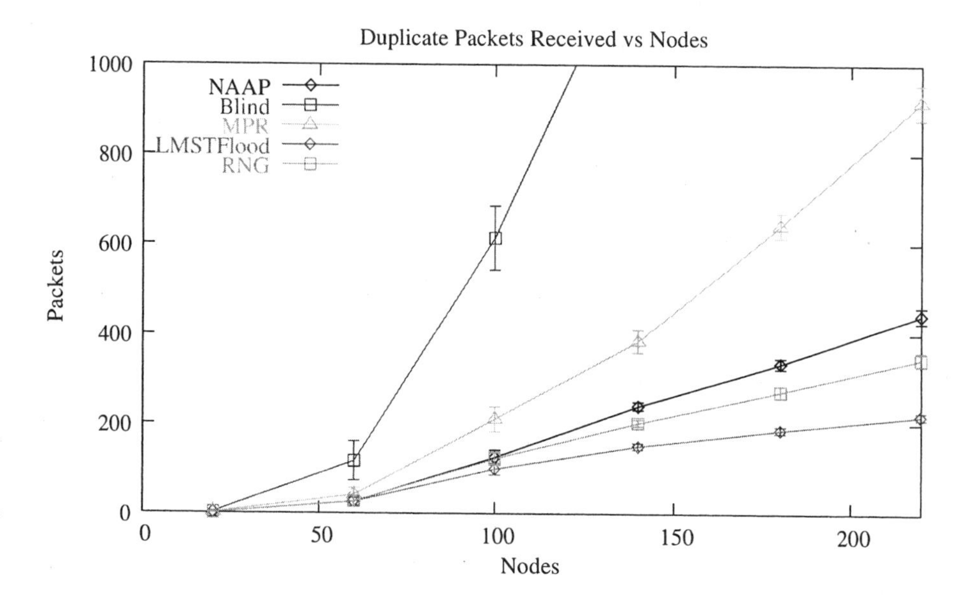

Figure 2-8. A comparison of duplicate packets recieved by NAAP, MPR, LMST-Flood, RNG and Blind flooding.

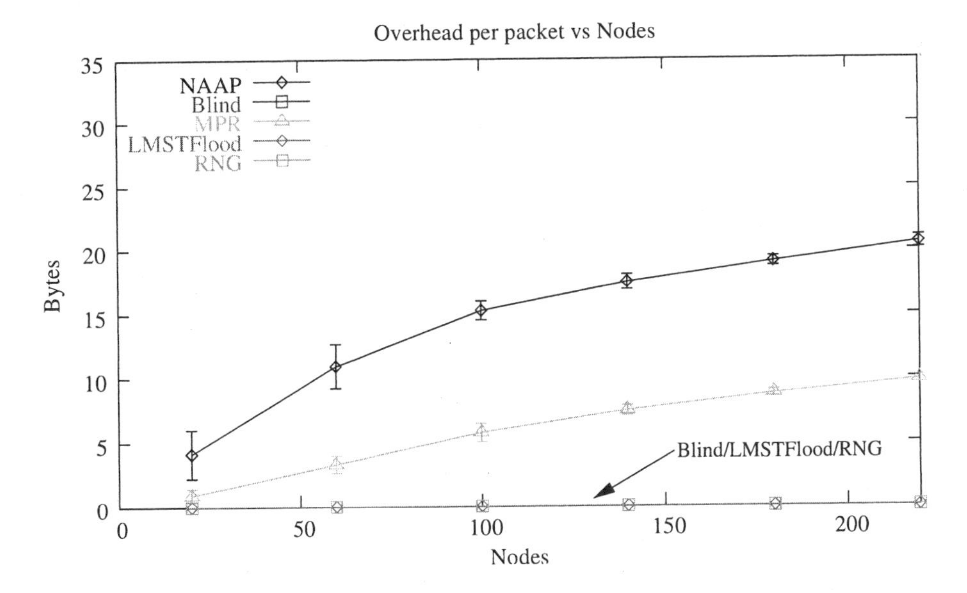

Figure 2-9. A comparison of the per packet overhead for NAAP, MPR, LMSTFlood, RNG and Blind flooding.

quickly as the other mechanisms. This is also evident by the resulting network radius of each mechanism as shown in Fig. 2-10.

Figure 2-9 shows average overhead per broadcast packet in bytes incurred by NAAP and MPR as the relay set is attached to each broadcast packet. In MPR the multipoint relay set may be distributed through beacon messages and therefore incurs overhead in beacon messages. In the simulation, we append the multipoint relay set to each packet prior to broadcast as done with source based MPR mechanisms [15]. RNG and LMSTFlood incure no additional overhead as each mechanism can determine independently whether or not to rebroadcast. It can be seen that as the node concentration increases, the required overhead of NAAP does not grow significantly. The calculation of overhead does not include neighbor discovery through beacon messages.

Figure 2-10 shows the resulting network radius in broadcast hops. Routing protocols may benefit from flooding mechanisms that have a lower network radius when performing route discovery. From Fig. 2-11 we can see that NAAP, RNG and LMSTFlood reduce transmission distance as the node density increases, therefore the network radius will increase for these mechanisms. MPR does not reduce broadcast power nor

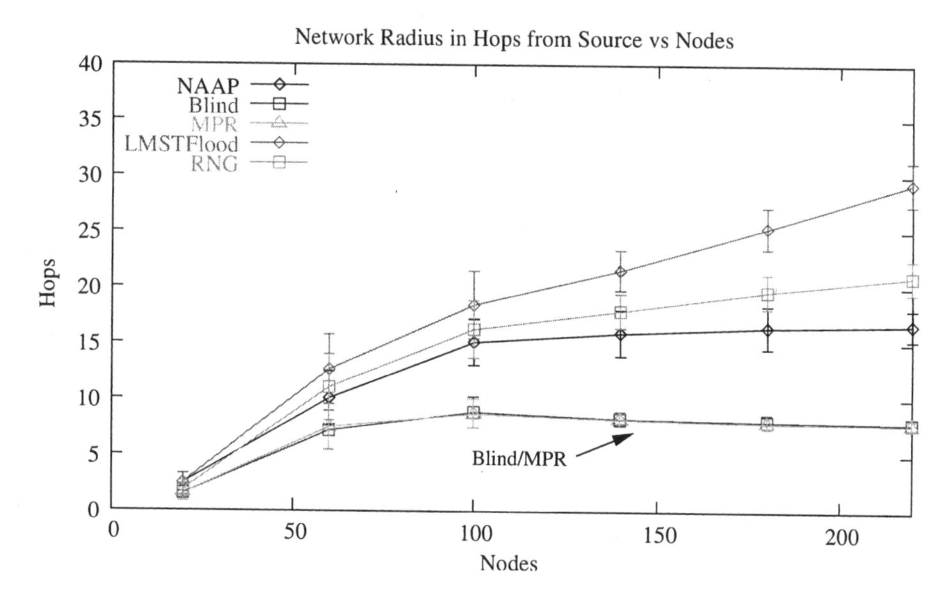

Figure 2-10. A comparison of the resulting network radius in hops for NAAP, MPR, LMSTFlood, RNG and Blind flooding.

introduce additional hops and therefore has the lowest network radius. MPR and NAAP may be the most useful to routing protocols as the network radius does not increase dramatically with an increase in node density. Allthough the network radius is higher than MPR, RNG and NAAP do have the added benefits of reducing the the broadcast storm problem more significantly than MPR, thereby providing improved performance.

Figure 2-12 shows the node coverage per broadcast with increasing density. As above we see that MPR does not reduce transmission distance and therefore as the node density increases more nodes are covered per broadcast, however as shown in Fig. 2-8 this results in significant duplicate packet reception. NAAP, RNG and LMSTFlood are able to restrict broadcast coverage as node density increases. Therefore they tend to be more scalable in higher node densities.

MPR has the lowest network radius, however it does not scale well to high node densities as it does not perform tranmission power control. The low network radius may be beneficial in routing protocols where less delay is required, however this comes at the price of high overhead in terms of the broadcast storm problem. Additionally MPR consumes

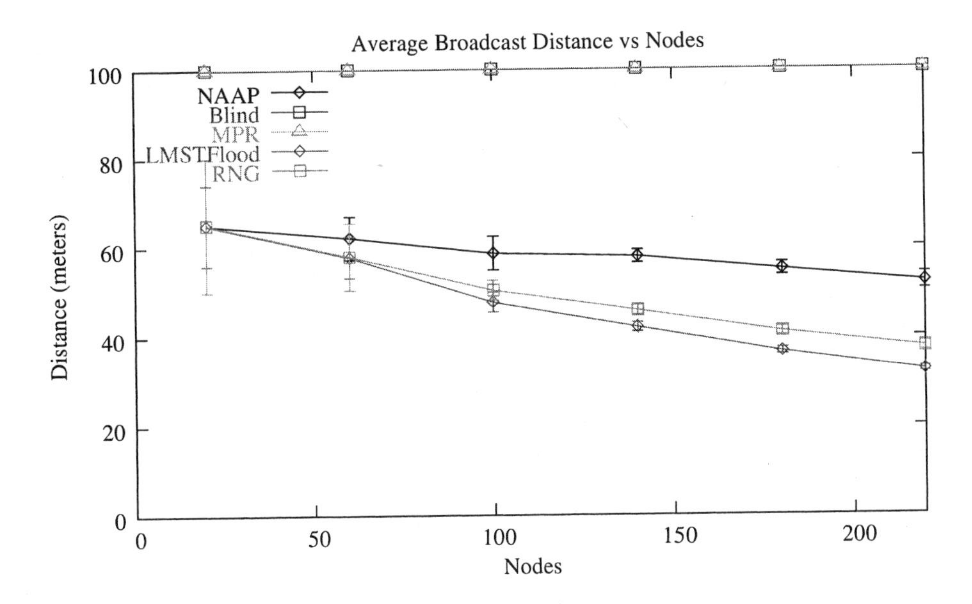

Figure 2-11. A comparison of the resulting average transmission distance for NAAP, MPR, LMSTFlood, RNG and Blind flooding

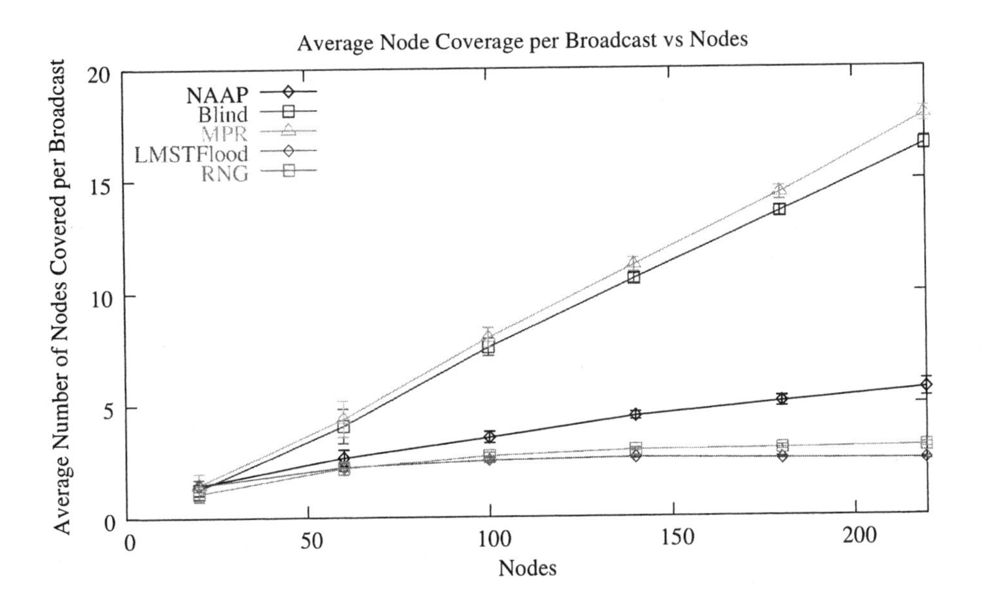

Figure 2-12. A comparison of the average node coverage per broadcast for NAAP, MPR, LMSTFlood, RNG and Blind flooding

more energy per broadcast than the other mechansisms. LMSTFlood provides the lowest overhead, lowest energy consumption, but highest network radius and may not be suitable to route discovery, but may be suited to disseminating link state information or network management information. The NAAP and RNG mechanisms show performance that ranges between LMSTFlood (lower bound) and MPR (upper bound). They may therefore be equally suitable to routing protocols or as general information dissemination mechanisms.

5. FLOODING RELIABILITY

In IEEE 802.11 [14], the basic medium access mechanism implemented at the MAC layer is Carrier Sense Multiple Access / Collision Avoidance (CSMA/CA). A node utilising CSMA with a data frame to transmit will first sense the shared medium by listening for any existing transmissions. If the medium is busy, then the station will delay its transmission. However, if the medium is not busy, then the node will begin transmitting the data frame. CSMA mechanisms are useful in scenarios where the shared medium is lightly loaded (low traffic) as there is minimal delay prior to transmission. The problem with CSMA is that if the shared medium is heavily loaded (high traffic), then the probability of nodes simultaneously sensing the medium as being free and then transmitting increases. Thus the possibility of obtaining collisions increases.

IEEE 802.11 provides two mechanisms for data transmission to one or more neighbouring nodes: unicast (one to one) and broadcast (one to all). It is important to note that significant differences exist between unicast and broadcast data transmission in IEEE 802.11. Unicast transmission allows a source node to send data directly to one destination node within transmission range. The MAC layer utilises CSMA/CA with Request To Send (RTS), Clear To Send (CTS) and positive acknowledgements. If a frame is not received or the CRC fails verification, then a retransmission will occur as no positive acknowledgement is received. The use of CSMA/CA, RTS/CTS and positive acknowledgements allows unicast transmission to be less susceptable to collisions, packet loss and the hidden node problem. Broadcast transmission allows a source node to send data directly to all nodes within transmission range. The medium access mechanism, CSMA/CA described earlier is not used in its entirety. Prior to data transmission, carrier sensing (CS) is performed by the MAC. If the medium is free then the MAC will transmit the data frame. If the medium is not free, then random backoff delay occurs. The MAC layer does not exchange RTS/CTS frames with surrounding nodes prior to transmission. The received data frame's CRC is verified upon reception at each receiver and allowed to progress up the protocol stack. If the data frame fails CRC verification, it is deleted. Unlike unicast transmission, no positive acknowledgement is transmitted back to the source. The broadcasting node has no mechanism to determine if a broadcast was received by one or all nodes. Given the lack of acknowledgements in broadcast transmission, no data frame retranmissions occur.

The use of the distributed MST algorithm in LMSTFlood allows for a highly optimised flood with the addition of some fault tolerance to improve reliability. However, there still exists a problem in broadcast environments where a broadcast packet may be lost due to noise corruption, packet collisions or hidden node transmissions. In LMSTFlood the number of neighbours that a broadcasting node may need to rebroadcast to is less than 1.5 neighbours on average, once the preceeding broadcasting node is removed from the BSET. There are therefore many situations where a packet may be lost and a flood may not propagate due to low fault tolerance. However, the size of the BSET (1.5 neighbours on average) allows for broadcast packet transmissions as used by all flooding mechanisms to be replaced by more reliable unicast packet transmissions such as that used in IEEE 802.11. Unicast transmissions are not completely reliable and packet loss is still possible as each packet will only be retransmitted at most a certain number of times. However, unicast provides a more reliable transport mechanism than broadcasting and requires no modifications to the MAC layer. The use of unicast transmission (as opposed to broadcast transmission) combined with LMSTFlood allows for a high degree of optimisation (given the broadcast storm problem) and increased reliability through an acceptable increase in redundancy and the addition of more reliable packet transmission.

6. CONCLUSIONS

In ad hoc networks the process of disseminating information throughout the network forms the basis of routing protocols, network management, service discovery and information collection. As ad hoc networks are broadcast in nature, it is important that this dissemination be done with minimal effect to the network. In this chapter, we compare the performance of existing distributed ad hoc network flooding mechanisms. Additionally, We propose to apply the Minimum Spanning Tree (MST) mechanism in a distributed manner with one hop topology information as the basis of a scaleable optimised flooding mechanism called Localised Minimum Spanning Tree Flooding (LMSTFlood). LMSTFlood is seen as a general information dissemination mechanism useful in high node concentrations. However due to the resulting high network radius it is not particularly suited to routing protocols. LMSTFlood significantly reduces energy consumption, utilises a smaller average transmission range and results in nodes receiving less duplicate packets during a flood. It is thus more effective at limiting the broadcast storm problem than existing optimised flooding mechanisms. In addition to the benefits of flooding optimisation, LMSTFlood combined with unicast packet transmission may be used to improve the overall reliability of a flood.

REFERENCES

1. S.Y. Ni, Y. C. Tseng, Y.S. Chen, and J.P. Sheu, "The Broadcast Storm Problem in a Mobile Ad hoc Networks". In *Proceedings of the Fifth Annual ACM/IEEE International Conference on Mobile Computing and Networking*, pages 151–162. ACM Press, 1999.
2. G. Toussaint, "The Relative Neighbourhood Graph of Finite Planar Set". *Pattern Recognition*, pages vol. 12, no. 4, pp. 261–268, 1980.
3. N. Li, J. C. Hou, and L. Sha, "Design and Analysis of an MST-based Topology Control Algorithm". In *Proceedings of IEEE Infocom 2003*, 2003.
4. B. Williams and T. Camp. "Comparison of broadcasting techniques for mobile ad hoc networks". In *Proceedings of MOBIHOC*, June 9-11 2002.
5. C.E. Perkins and E.M. Royer, "Ad Hoc On-Demand Distance Vector (AODV) Routing". In *Proceedings of the Second Annual IEEE Workshop on Mobile Computing Systems and Applications*, pages 90–100, Febuary 1999.
6. D.B. Johnson, D.A. Maltz, and J. Broch, *"DSR: The Dynamic Source Routing Protocol for Multihop Wireless Ad Hoc Networks", in Ad Hoc Networking*, chapter 5, pages 139–172. Addison Wesley, 2001.
7. A. Qayyum, L. Viennot, and A. Laouiti, "Multipoint Relaying: An Efficient Technique for Flooding in Mobile Wireless Networks". In *Proceedings of 35th Annual Hawaii International Conference on System Sciences, 2001.*
8. P. Jacquet, P. Muhlethaler, A. Qayyum, A. Laouitim, and L. Viennot, "Optimized link state routing". draft-ietf-manet-olsr-06.txt, 2000.
9. J. Lipman, P. Boustead, and J. Judge, "Neighbor Aware Adaptive Power Flooding in Mobile Ad hoc Networks". *International Journal of Foundations of Computer Science Vol. 14, No. 2*, 14(2):237–252, April 2003.
10. S.A. Borbash and E.H. Jennings, "Distributed Topology Control Algorithm for Multihop Wireless Networks". In *Proceedings 2002 World Congress on Computational Intelligence (WCCI 2002)*, Honolulu, Hawaii, 2002.
11. J. Cartigny, F. Ingelrest, and D. Simplot, "RNG Relay Subset Flooding Protocols in Mobile Ad hoc Networks." In *International Journal of Foundations of Computer Science Vol. 14, No. 2*, pages 253–265, April 2003.
12. R. Prim, "Shortest Connection Networks and some Generalisations". *The Bell System Technical Journal*, pages vol. 36, pp. 1389–1401, 1957.

13. W.R. Heinzelman, A. Chandrakasan, and H. Balakrishnan, "Energy-efficient Communication Protocol for Wireless Microsensor Networks". In *Proceedings of the Hawaii International Conference on System Sciences*, pages 1–10, January 2000.

14. IEEE, "IEEE Std. 802.11-1997 Wireless LAN Medium Access Control (MAC) and Physical Layer (PHY) Specification". http://standards.ieee.org/getieee802/, 1997.

15. J. Lipman, P. Boustead, J. Chicharo, and J. Judge, "Resource Aware Information Dissemination in Ad hoc Networks". In *Proceedings of the 11th IEEE International Conference on Networks (ICON 2003)*, Sydney, Australia, September 2003.

Chapter 3

A PRESENCE SYSTEM FOR AUTONOMOUS NETWORKS

Anthony Dang and Bjorn Landfeld

Advanced Networking Research Group, School of Information Technologies, The University of Sydney, Australia

Abstract: This chapter addresses the increasingly important issue of effective addressing, object location and presence notification in networks with no infrastructure and dynamically changing environments. This chapter presents Calto, an architecture comprising presence concepts from Second and Third Generation Mobile networks, Instant Messaging Systems subscriber services, and distributed DNS style functionality in a Peer-to-Peer setting. Calto utilizes locality and key nodes to provide these services. The architecture accommodates for true Ad Hoc environments while being scalable and robust in the face of network instability.

Key words: Ad Hoc, Peer to Peer, Resource and Service discovery, Presence, Location, Searching, Distributed Networks

1. INTRODUCTION

Networking environments are continuously evolving. The more recent introduction of wireless networks such as cellular networks [1], wireless LAN [2] and Personal Area Networks, and the possible interconnections between these networks (see Fig. 3-1) have created a new set of challenges [3,4,5]. The relatively new paradigm of Ad Hoc networks [6] introduces dynamically changing topologies (nodes can join, leave and move) that can be formed without the need of fixed infrastructure. Furthermore, with the recent strides in wireless technologies we now have the case where entire networks may gain and loose connectivity, that is, networks can merge and partition.

In a centralized network, nodes advertise their availability and current location to the centralized server. Services such as object location, DNS and Instant Messaging (IM) [7] are easily provided as centralized servers are generally consistently available and do not change in location. However, with dynamically changing networks, this task becomes much more challenging.

Recently there has been much research in regard to efficient network layer routing to nodes in the wireless domain. Research into effective addressing, presence notification [7] and location in these networks has so far been lagging behind.

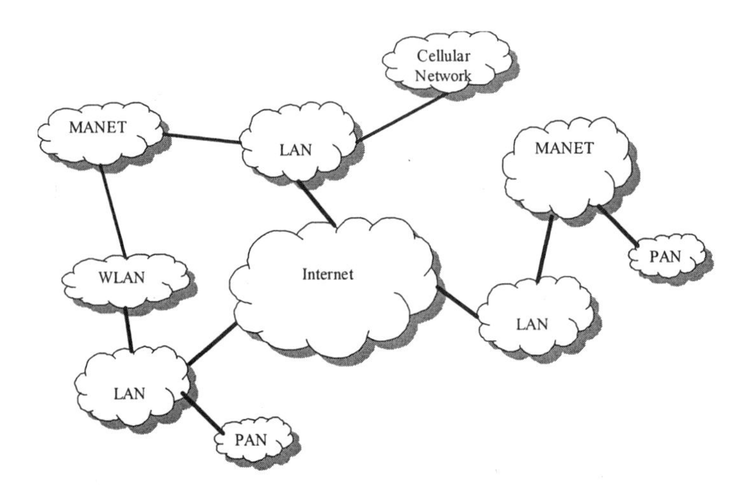

Figure 3-1. Converging networking technologies

In this chapter, we introduce Calto; A Self Sufficient Presence System for Autonomous Networks. In order to provide a presence service in an Ad Hoc environment, Calto integrates GSM [1] and Peer-to-Peer (P2P) concepts [8,9], a distributed lookup architecture [12,13,14], subscriber services [7] and a specialized search algorithm. This combination makes up a truly dynamic system in the face of changing network topologies. We show that Calto scales well and thus can support a large number of nodes without introducing significant overhead. This chapter is organized as follows. In section 2 we describe the technologies that constitute the system. In section 3, we describe the Calto framework. Section 4 explains the presence information distribution protocols. Section 5 describes the search algorithm. In section 6 we describe how Calto networks are formed and updated. In section 7 we provide an analysis of the scalability of the system before presenting future work in section 8 and finally concluding in section 9.

2. BACKGROUND

Advances in link layer technologies such as IEEE 802.11 [2] and Bluetooth [10], have allowed for the wireless scene to be extended out of the domain of centralized systems and into the realm of Ad Hoc networks. These networks are envisioned to operate in unstable network environments and without the need for centralized servers. Tracking objects in such a system is inherently difficult. The advent of the Peer-to-Peer paradigm (application layer overlay networks [8,9], communicating via direct links between nodes) has made these hopes attainable. In this section we discuss the applicability of these technologies to this area of research.

2.1 Wireless Location Tracking in GSM

There are many challenges and issues that effect the development of mobile computing [5]. Of particular interest for this research are the problems associated with object location.

GSM cellular networks implement a two-tier scheme where a home database, termed Home Location Register (HLR), is associated with each mobile user. The HLR maintains the current location of the user, that is, a zone in the cellular network where they are currently located. GSM calls this a Visiting Location Register (VLR). To locate a user, their HLR is identified and queried. When a user moves to a new zone, their HLR is contacted and the new location is stored. In GSM systems this method has shown much promise in the way of presence and location of mobile objects.

2.2 Peer-to-Peer Networks

Peer-to-peer networks appeared on the scene a few years ago in the form of file sharing networks. Rather than downloading files from a centralized server, nodes in the network downloaded them from their peers (hence the name!). The centralized server was used only to index which nodes in the network had what files. Later models of the paradigm completely eliminated the need for a centralized server [11].

The advent of the P2P paradigm has shown many benefits in the area of robustness of distributed systems. Distributed peer-to-peer lookup architectures such as Tapestry [12], CAN [13] and Chord [14] have been proposed for the efficient storage and searching of objects in large scale distributed P2P systems. Unfortunately these proposals do not allow for parts to merge, separate and re-attach, as would be the case in an Ad Hoc environment.

This notion of location, notification and presence in an Ad Hoc network is currently an unexplored area. In our work, we use concepts of the above technologies to achieve this functionality and provide a system for locating users, nodes and services in highly dynamic wired and wireless environments.

3. CALTO

Calto models some of its aspects on Chord as its simplistic design enables various enhancements that enable Calto's functionality. In this section we present Calto's information storage structure, efficiency enhancements and the protocols that enable adaptive addressing.

3.1 Basic Structure

Calto operates as a P2P application-layer overlay network organized in a ring structure, where each node is ordered in terms of an identifier. The architecture provides a distributed lookup service, allowing the insertion, updating, lookup, and deletion of key-value pairs using identifiers as handles to the keys. Examples of searchable identifiers are email addresses, IP addresses, a string name of a service, or even a file name. The node identifiers utilize locality (see section 3.3) and are generated by applying an Ordered Hashing function to the searchable identifier (the identifier known to the person or process conducting a search).

Nodes in Calto are responsible for identifiers that are numerically close to the node's own identifier. For example, in Fig. 3-2 we can see that the identifiers 1 and 2 are numerically closer to 1 than to 4, so their assignments are to node 1. Node 0 is a special case. The example in Fig. 3-2 depicts a small namespace from 0 to 7. Node 0 is the keeper of identifier 7 because it is mathematically closest to it (closer than node 4!) when the namespace is wrapped around in a ring (or circular list).

In Fig. 3-2, it can be seen that any insertion of a key-value pair involves, simply, performing a search for a node that has an identifier close to the identifier of the pair, and then inserting at that node. Mirrored copies of node's tables are kept in r consecutive nodes preceding and succeeding a node in the ring. So, when a node leaves a network, the responsibility for its keys is implicitly distributed between its immediate successor and predecessor.

It should be noted that this system can store presence information (identifiers) about anything with a unique identifier. This unique identifier can be hashed into a Calto identifier and stored in the system with a mapping to the address or identifier of the node on which this service, resource etc resides.

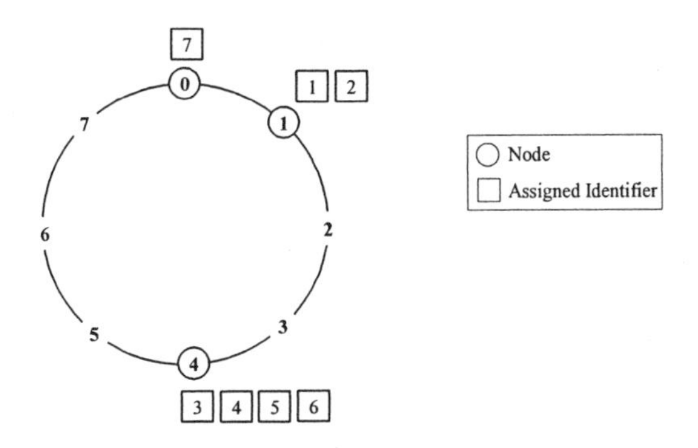

Figure 3-2. Calto identifier ring with a namespace of 0-7

3.2 A Multi-tier Architecture

In any network (large or small) nodes may differ in resource availability (storage capacity, processing power, bandwidth, online availability etc). Calto endeavors to utilize these differences in a hierarchical P2P design, providing scope and flexibility for hybrid [8] client/server functionality (when available) to improve network performance. Specifically, nodes can be elected to be *Super Peers*, providing a similar function to a server. These nodes hold copies of key-value pairs of their children and adjacent *Super Peers* allowing for faster searching.

Super Peers maintain their own logical ring overlaying the Global Ring (see Fig. 3-3), enabling the possibility of information backup, and increased performance for information retrieval. The advantage of this design is that it also leaves the network open for the utilization of nodes that may indeed be high-powered centralized servers (permanent or otherwise).

3.3 Exploiting Locality

Locality is important when dealing with joined networks. It is undesirable to pass messages through another network if queries can be resolved locally since it both introduces overhead traffic and drains battery resources. Utilizing the physical layer efficiently is particularly important in this context.

In Chord, all the nodes in the network are organized in a global one-tier ring, and are inherently randomly distributed. In Calto we wish to make use of locality, so rather than randomly distributing identifiers, we use a node's IP address and spread it evenly over a 160bit address space. For added uniqueness assurance we also add a random value.

Calto endeavors to utilize physical locality to increase network performance, and reduce the cost of sending messages across the physical and data link layers. Thus, Calto allows multiple logical (locality based) rings to overlay its Global Ring structure. The aim is to allow nodes to function independently in Local Rings in addition to operating in a Global Ring context. In other words, we wish to allow nodes to gain access to information about the presence of physically close nodes by using the physical layer as little as possible. In Fig. 3-3 we show a possible network formed by 3 smaller networks (note the dotted lines), all of which consisting of *Super Peers* which communicate on their own tier.

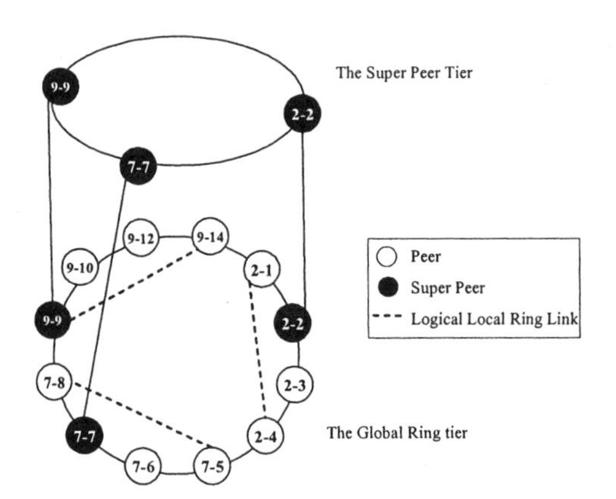

Figure 3-3. Local and Global Ring overlays and the *Super Peer* Tier

4. PRESENCE PROTOCOLS

In a dynamically changing network we wish to know (be notified) when an object becomes available (its presence), and where it currently is (its location). In this section we describe how this can be achieved using concepts from GSM systems and Subscriber Services.

4.1 GSM Style Presence Keepers

In a network where objects can change location, it is important to be able to find them again efficiently. Obviously, if an object never changes location then a search using the original identifier may be performed. However, since we are dealing with the situation where objects can change location then we must have a way of finding them without having to flood the network.

In GSM cellular networks a node will be assigned to Home Location Register (HLR), which will store the current location in the network of that node. The notion of a HLR in Calto is a construct assigned to any node on the network. In effect, in Calto, a HLR may join and leave the network at any time. The nodes a HLR serves are then implicitly assigned to one of a set of possible relative backup nodes to the HLR, as a search would result in one of these nodes, as follows.

When the node enters the system it inserts an identifier in the form "nodeID-fn('HLR')" where nodeID is the node's identifier, fn() is a hash function, "HLR" is a string and the value of the pair is the node's current IP address. This allows a node to change location, as its HLR will always be a node close to the identifier "nodeID" which holds a mapping to its current address. A search for a node is simply a search for its HLR.

4.2 Subscribing to Presence

To be notified of the availability and presence of other nodes and objects in the system, a node must subscribe to the information. Thus each node in the network is aware of the nodes that have subscribed to its presence.

When a node comes online, it can deposit a presence flag in the network for all nodes that have subscribed to its presence, but are not currently online. We call the node that holds flag a *Presence Box* (analogous to an email box). The flag is similar to the HLR insertion and is in the form "$\text{nodeID}_{\text{sub}}$-fn('Presence Box')", where $\text{nodeID}_{\text{sub}}$ is the identifier of the subscriber. The keeper (and relative backup nodes) of this identifier would merely concatenate this new flag to a list. When a node comes online it searches for "nodeID-fn('Presence Box')" (where nodeID is its identifier). Note that this state information is not by any means the only type of information Calto can relay. Any information that can be associated with a presence service can be stored in the Presence Box.

4.3 Adaptive Addressing

In a dynamic environment nodes will join, leave and also change location. Information stored in these nodes still needs to be accessible. Without an adaptive and ordered addressing scheme the only mechanism to find this information is by means of flooding.

In flooding, a node will send a message to all of its neighbors. All the receiving nodes then send copies of the message to all of their neighbors, excluding the node it received the message from. This is a very effective method of reaching all the nodes on the network. However, we have the case where many nodes will receive the same message more than once, and therefore flooding (although effective) can lead to a massive waste of resources [15].

In Calto when a node leaves and rejoins the network in a different location (with different IP address), a session identifier (based on its new IP address) is used to place the node in the network, enabling the exploitation of locality when nodes move. Searching for a node merely involves searching for its original identifier, which will map to its current location (IP address).

To distinguish between *Local Rings* each node has the identifier of the initial *Super Peer* as a prefix to their own. This prefix acts as a domain name, and if the *Super Peer* is indeed a permanent centralized server, then this scheme compliments the current domain scheme of the Global Internet, and can be easily integrated.

5. SEARCHING

Efficient and fast searching is always desirable, and it is paramount in systems that are limited in resources such as in battery constrained mobile computing. Searches must therefore have a predictable upper bound and use the physical layer as little as possible.

Calto's search algorithm is similar to that of Chord's, as each node stores information about only a small number of other nodes in the network. In a stable network the search requires only $\log_2 N$ messages (where N is the number of nodes in the network). This scalability is achieved via specific routing tables for each node.

It should be noted that Calto is not merely limited to searching for nodes. The system is capable of finding email addresses, IP addresses, files, and any string name of a service, resource, device or person. For example, we can search for a person on the network by using their known email address.

We firstly search for a node in the network that has an identifier that is numerically close to "fn('email')", where email is the email address of the desired person. This identifier will map to the desired person's nodeID(s). The next step is to search for "nodeID-fn('HLR')" to get the current IP address. The same will apply to finding files on the network. The identifier "fn('filename')" can be used to find the identifiers of the node(s) that hold this file.

5.1 Routing Tables

Each node in Calto requires a selection of nodes based on exponentially increasing distance (node hops) in both the clockwise and anti-clockwise directions. Specifically, each node requires in its routing table the nodes that are 2^0, 2^1, 2^2, 2^3,..., $2^{\log N-1}$ hops from its current position in the ring, where N is the number of nodes in the network. These routing entries allow for a search to occur in under $\log_2 N$ node hops.

Note that in Chord, nodes only hold routing entries in the clockwise direction. In Calto we wish to accommodate for network partitioning and independent operation of partitioned networks. In order to achieve this in Calto, routing entries are for both clockwise and anticlockwise directions (see Fig. 3-5 (a)). This is necessary since Ad Hoc environments are highly dynamic and segmentation is likely to occur at times. Therefore by allowing nodes to know about other nodes in both directions around the ring, there is a higher chance of either repairing the ring, or allowing the formation of individual rings from the broken off sections.

5.2 Search Algorithm

When a node initiates a query it will search its routing tables (its predecessor list *p* and its successor list *s*) for a node with an identifier that is numerically closest to the search identifier. It then passes the query to that node, and the query is performed recursively until it is routed to the node that contains the key-value pair the identifier belongs to (see Fig. 3-4).

It should be noted that although these routes may overshoot the node that is responsible for the queried identifier, the namespace is significantly reduced at each step (see Fig. 3-5 (b)). In fact, at each hop we reduce the remaining number of possibilities by half in each direction. Therefore, we have a search time of $\log_2 N$, where N is the number of nodes within the network.

As an optimization, if the node is a member of a LAN, before it routes a query, it may perform a Local broadcast check for any nodes on its LAN that are not in its routing table which might have an identifier closer to the target than its own identifier.

```
search(id)
if n = closest(id)
  n.search(id)
else
  return result

closest(id)
for i = 0 up to s.size-1
  if s_i - id < s_i+1 – id
    n1 = s_i
for i = 0 up to p.size-1
  if p_i - id < p_i+1 – id
    n2 = pi
if n1 – id < n2 – id
  return n1
else
  return n2
```

Figure 3-4. Search algorithm

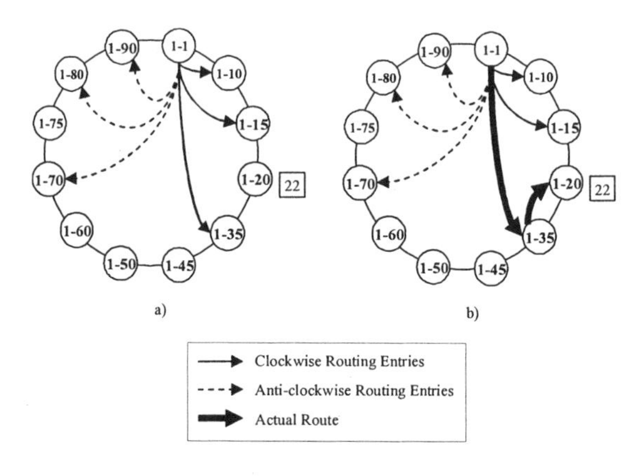

Figure 3-5. Calto's search

5.3 Searching Connected Networks

We mentioned previously that a search requires only $\log_2 N$ messages. In the scenario where we have connected networks this is not the case. We have the situation where each smaller network has nodes in their routing tables that are only local to their own networks. So we assert that the maximum number of messages required is the sum of $\log_2 n_i$. That is the number of messages to find a route around the ring to the next network.

6. CALTO NETWORK FORMATION

6.1 Node Insertion

Assuming a node can contact any other node in the network, it will be able to find the place in the ring to insert itself through a search. This may or may not involve directly contacting a *Super Peer*. The procedure for the new node is to contact the nodes that will be its successor and predecessor in the ring, and have them populate the new node's routing tables.

A new node Y needs to know about nodes that are $2^0, 2^1, 2^2, 2^3, \ldots, 2^{\log N-1}$ hops away from it. If Y is inserted between nodes X and Z, then Y will take on half the routing entries of these two nodes (see Fig. 3-6). That is, node Z (the immediate successor in the clockwise direction) will know about the nodes that are needed in the anti-clockwise direction (and vice versa for node X). Therefore, all that is needed is for the new node Y to contact these nodes to update their routing entries to point to Y rather than X and Z.

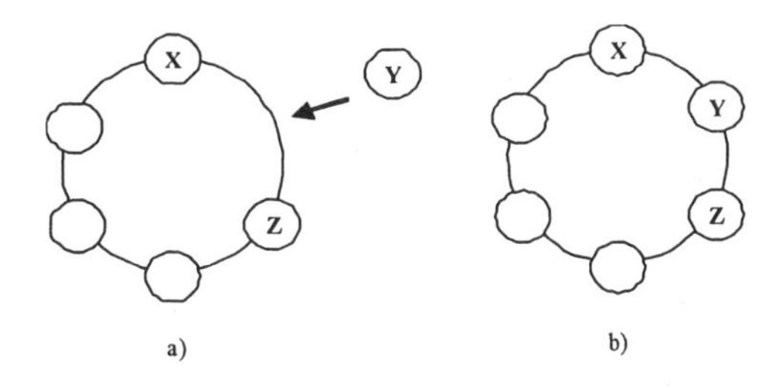

Figure 3-6. Node insertion.

This procedure allows for $2\log_2 N$ nodes to have their routing tables updated, reflecting the presence of the new node. It should be noted that due to the associated overhead in updating large dynamically changing networks, routing entries would only be able to approximate the desired node hop distribution. However, scalable routing is still possible with routing approximations.

6.2 Routing Table Updating

In a changing topology, routing tables will become out of date and periodic updating is therefore needed. Each routing entry in Calto requires nodes that are exponential hops away ($2^0, 2^1, 2^2, 2^3, \ldots, 2^{\log N-1}$ hops).

For a node to update its routing tables it requires help from other nodes. A node will ask its 2^{ith} node in its routing table for its 2^{ith} node in the same direction. The returned node is in fact the calling node's 2^{i+1th} entry. Note that a correct routing entry for the calling node requires that the queried node has a correct entry to return. So we assert that each node in the network must know its immediate predecessor and immediate successor. Fig. 3-7 shows this algorithm for updating a node's successor list s.

```
refresh()
for i = 0 up to ∞
  request s_i from s_i // s_i of s_i is s_{i+1} for this node
  if no result or this.id < result > s_i
    stop
  else
    add/replace s_{i+1} with result
```

Figure 3-7. Refresh routing table algorithm.

6.3 Node Departures and Network Partitioning

In Calto when a key-value insertion is made, the pair is inserted at r consecutive nodes on both sides of the actual keeper of the pair (The quantity r is a configuration parameter that depends on the degree of redundancy required). When a node leaves the network its successor and predecessor implicitly take on the responsibility of the key-value pairs that are closest to their identifiers.

To function in a dynamically changing environment, Bluetooth allows *Piconets* to form and merge independently and function autonomously when a partition occurs. Calto also adapts to these situations, allowing for previously formed LANs, WANs or Mobile Ad Hoc Networks to merge and partition. Calto allows for its addressing scheme to operate even when networks become partitioned.

When a node notices that its immediate predecessor or successor is missing then it can simply check other nodes in its routing tables to find the closest contactable node to use as its immediate predecessor or successor.

This algorithm is also used for when a network becomes partitioned. The ring healing process will result in a partitioned network forming smaller rings. Essentially, the ring is split and what remain are 2 or more arcs of nodes (see Fig. 3-8). The nodes at the ends of each broken arc can search for each other and then form independent (smaller) rings. See Fig. 3-9 for the algorithm.

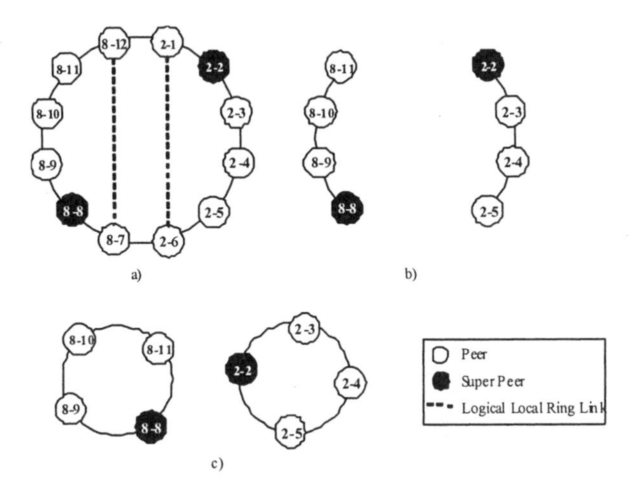

Figure 3-8. Network Partitioning

6.4 Network Merging

When two nodes from different networks are able to contact each other we logically break each ring at their start and end points (lowest and highest identifier found by search or *Super Peer* query), and attach each start point to the end point of the other broken ring, forming a larger ring. In Fig. 3-10 (a), (b) and (c) we can see this progression of joining two networks. As can be seen in Fig. 3-10 (b) and (c) we maintain the Local Ring connections (dotted lines) after the merger into the larger ring. This is to allow the original rings to function, taking advantage of locality. In addition, the *Super Peers* of the merging rings also get inserted in the correct places on the Super Peer tier, as can be seen in Fig. 3-10 (d).

```
detect_and_heal()
if ping(p_0) == no response
  remove(p_0)
while (n = find_nearest_contactable_p() != no result)
  p = reverse(n.s)  // use as p, s from n in reverse order
  p.add_last(n)
   for i = 1 up to p.size-1
     if p_i > this.id
       remove(p_i)


find_nearest_contactable_p()
for i = 1 up to p.size-1
  if ping(p_i) == no response
     remove(p_i)
  else
     return pi
for i = s.size-1 up to 0
  if ping(s_i) == no response
     remove(s_i)
  else
     return p_i
return no result
```

Figure 3-9. Ring Healing and Network Partitioning Algorithm

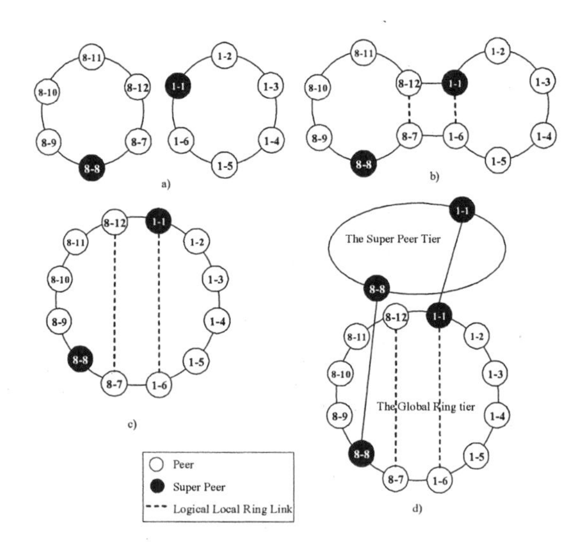

Figure 3-10. Merging Networks

7. ANALYSIS

In this section we give an analysis of the scalability of Calto's searching algorithm and give an example of network traffic introduced by node insertion.

7.1 Searching

Calto's search algorithm scalability performs with an upper bound of $O(log_2 N)$ node hops where N is the number of nodes in the network. The convergence distance at each step in the algorithm to the target identifier holds true for the following relation:

$$|k_i| < |k_{i-1}/2|,$$

where k is the converging distance to the target identifier.

As mentioned previously, each node in the network holds in its routing table nodes that are $2^0, 2^1, 2^2, 2^3, \ldots, 2^{\log N-1}$ hops away. We assert that at every step i, in the algorithm, the maximum node hop 2^l can be at most $2^{(\log N-1)-i}$ where $2^{\log N-1}$ is the largest leap in the routing table. Specifically at each step, it is as if we removed 2^l_{i-1} from the possible set of left over routes in all the node's routing tables. That is, the distance between the node and 2^l_i hops away is equal to the distance between the nodes at 2^l_i and 2^l_{i-1} in the routing table. Therefore, for every 2^l_{i-1} we remove from the possible set of next routes, we halve the remaining number of nodes to search. Thus, the convergence is linear with an upper bound of $O(\log_2 N)$, and an average of $(1/2\log_2 N)$.

7.2 Node Arrival Traffic Analysis

In this section we will give an example and show that the amount of traffic introduced for a node insertion into the network is quite minimal and acceptable.

Every node has a successor and predecessor routing table with of $O(\log_2 N)$ entries each. When a node joins the network, there must be $O(2\log_2 N)$ messages to initialize its routing table, and to update the routing tables of others. To provide an example of an actual traffic load we give the following example. Assuming an implementation of IPv6 (128bit IP address), a Calto identifier size of 160bit, and accounting for various bit flags that may be needed in the implementation, we have a packet size of approximately 300bit.

In a network of size 1,000 nodes ($N = 1{,}000$), the amount of traffic on the network would be only 6Bytes ($300 \times 2\log_2 1000$). It should be noted that in a dynamically changing network we might have the case where a node's routing tables are unable to be updated fast enough to reflect the true topology of the network. Scalable routing is still possible as long as the routing entries still approximate the distances in terms of hops.

8. FUTURE WORK

Application-layer routing can sometimes utilize the physical layer poorly. Much research needs to be done in regard to applying application-layer routing in conformance with network layer and link layer routing protocols. This is especially important in power-constrained networks such as Mobile Ad Hoc Networks.

Calto allows the selection of *Super Peers* in the network. The selection criteria for electing these nodes will vary for each situation. For example, in a wireless environment we have limited bandwidth, varying energy resources and varying levels of mobility. There is a trade-off between these resource constraints that is currently not fully understood and needs to be investigated.

In the wired domain nodes with high bandwidth potential would be ideal as *Super Peers* as any request would be quickly responded to. On the other hand, if a node has statistically proven to be very sparse in its online availability, then it would not be a good choice to elect it as a *Super Peer*. The trade-off here is in terms of bandwidth and online availability.

9. CONCLUSION

In this chapter we have presented inherent problems with the convergence of new networking technologies. We have discussed the difficulty of applying presence, notification and searching of traditional network models in an Ad Hoc network environment, and have shown that there needs to be much research into this area if effective convergence of these technologies is to be realized. To date, there has been little research in this area, and we have presented Calto as a possible solution to these concerns.

Calto's design allows for application layer overlay networks to form, merge and partition, and is a networking architecture designed for all environments (wired or wireless). Calto models some of its fundamental attributes (such as its circular topology and search) on the Chord architecture, but there are some fundamental differences. Chord prefers to randomly place nodes in its overlay ring, where Calto endeavors to keep local nodes close together to benefit from topological proximity.

Calto is aimed at providing an architecture that would be portable and dynamic between all domains, wired and wireless, with scarce or plentiful resources. Utilizing locality is clearly beneficial to a wireless network as bandwidth and power are scarce, and also to a wired environment where LAN resources can be exploited.

Calto is to the best of our knowledge the first known attempt of implementing presence awareness in Ad Hoc Networks. Its unique integration of Peer-to-Peer design, distributed lookup architectures, subscriber services of IM systems, a specialized search algorithm and cellular network concepts makes it truly unique and dynamic in the face of changing topologies.

10. REFERENCES

1. Mouly, Michel, and Marie-Bernadette Pautet. ³The GSM System for Mobile Communications², Telecom Publishing, France June 1992, ISBN:0945592159
2. Wireless LAN Medium Access Control and Physical Layer Specification, IEEE Std. 802.11, 1999.
3. J.D. Solomon, Mobile IP: The Internet Unplugged, 1st ed., Prentice Hall, 1998. ISBN: 0138562466
4. Charles E. Perkins, ³Mobile Networking Through Mobile IP², IEEE Internet Computing, vol. 2, Issue 1, pp. 58-69, 1998, ISSN:1089-7801.
5. George. H. Forman and John Zahorjan, "The Challenges of Mobile Computing," *IEEE Computer*, 27(4):38–47, April 1994
6. Carlo Kopp, "Ad Hoc Networking", *Systems*, June 1999, pp33-40, http://www.csse.monash.edu.au/research/san/AdHocNetworks.pdf
7. D. Crocker, "Common Presence and Instant Messaging (CPIM)", IETF Internet Draft, 14-AUG-02. http://www.ietf.org/internet-drafts/draft-ietf-impp-cpim-03.txt
8. Rüdiger Schollmeier, "A Definition of Peer-to-Peer Networking for the Classification of Peer-to-Peer Architectures and Applications", *Proceedings of the IEEE 2001 International Conference on Peer-to-Peer Computing (P2P2001)*, Linköping, Sweden, August 27-29, 2001.
9. David G. Andersen, Hari Balakrishnan, M. Frans Kaashoek, Robert Morris, "Resilient Overlay Networks", *Proc. 18th ACM SOSP*, Banff, Canada, October 2001
10. Jaap C. Haartsen, "The Bluetooth Radio System," *IEEE Personal Communications*, Vol 7, No 1, pp 28-36, February 2000
11. *The Gnutella Protocol Specification v0.4* - http://www.clip2.com
12. B. Y. Zhao, J. D. Kubiatowicz, and A. D. Joseph. Tapestry: "An infrastructure for fault resilient wide-area location and routing," *Technical Report UCB//CSD-01-1141*, U. C. Berkeley, April 2001

13. S. Ratnasamy, P. Francis, M. Handley, R. Karp and S. Shenker. "A scalable content-addressable network," *Proceedings of the ACM SIGCOMM 2001,* San Diego, CA, Aug. 2001.
14. Stoica, R. Morris, D. Karger, F. Kaashoek, H. Balakrishnan. Chord: "A Scalable Peer-to-Peer Lookup Service for Internet Applications," *Proceedings ACM Sigcomm 2001*, San Diego, CA, Aug. 2001.
15. M.Portmann, A. Seneviratne "The Cost of Application-level Broadcast in a Fully Decentralized Peer-to-peer Network" *ISCC*, Sicily, Italy 2002.

Chapter 4

SECURE ROUTING PROTOCOLS FOR MOBILE AD-HOC WIRELESS NETWORKS

Asad Amir Pirzada and Chris McDonald
School of Computer Science & Software Engineering, The University of Western Australia, 35 Stirling Highway, Crawley, W.A. 6009, Australia.

Abstract: An ad-hoc network comprises of limited range wireless nodes that function in a cooperative manner so as to increase the overall range of the network. Each node in the network pledges to help its neighbours by passing packets to and fro in return of a similar assurance from them. All is well if the participating nodes uphold such an altruistic behaviour. However, this is not always the case and often nodes are subjected to a variety of attacks by other nodes. These attacks range from naive eavesdropping to vicious battery draining attacks. Routing protocols, data, bandwidth and battery power are the common targets of these attacks. In order to overcome such attacks a number of routing protocols have been devised that use cryptographic algorithms to secure the routing mechanism, which in turn protects the other likely targets. This chapter gives an overview of seven such secure routing protocols by presenting their characteristics and functionality, and then provides a comparison and discussion of their respective merits and drawbacks.

Key words: Security; Ad-hoc Wireless Networks; MANET; Routing Protocols.

1. INTRODUCTION

In wireless ad-hoc networks, the nodes act like mobile IP routers and carry out basic functions like packet forwarding, routing and network management. Due to a number of factors including dynamic topology, energy constraints and uni-directionality of links, standard intra-router protocols cannot be immediately applied to mobile ad-hoc wireless networks. In addition, as the routers are moving most of the time, the network lacks traffic concentration points a basic requirement of standard routing protocols. Many different types of routing protocols have been specifically developed for ad-hoc networks and have been classified into two categories as Reactive and Proactive [21]. In Reactive Routing Protocols, in order to preserve a node's battery, routes are only discovered when essentially required, while in Proactive Routing Protocols routes are established before use and hence avoid the latency delays incurred while discovering new routes. The effective operation of these routing protocols entails consistent benevolent behaviour from all participating nodes. This is more than often difficult to accomplish in open networks and requires some out-of-band mechanism for establishing and maintaining trust in the network [19]. Cryptographic mechanisms have been a major tool enforcing mutual trust relationships among the wireless nodes for the protection of routing protocols.

The transport medium being radio waves allows multiple recipients at the same time without any authenticity of either the sender or the receiver. The core of the security problems that are specific to ad-hoc networks exist because of this difference from wired networks. The nodes of an ad-hoc network cannot be trusted for the accurate execution of vital network functions, as opposed to the dedicated nodes of a traditional network. For example, a malicious node may try to redirect routes, advertise false routing information or engage a node in resource consuming activities such as routing packets in a loop. Furthermore, as the nodes of an ad-hoc network operate in a cooperative manner in a broadcast medium, ad-hoc wireless networks are more vulnerable to attacks.

Several routing protocols for ad-hoc networks have been developed to secure the routing process. This chapter examines these routing protocols by first describing the operation of each of these protocols and then comparing their various characteristics. The remainder of the chapter is organized as follows. In Section 2 we discuss some attacks that are carried out against ad-hoc networks. In Section 3 we examine the vulnerabilities, which are present in some common ad-hoc network routing protocols. Section 4 highlights some of the security issues related to ad-hoc networks. Section 5 gives an overview of recent work on secure routing protocols for ad-hoc networks. Section 6 discusses the comparison of various performance parameters of these routing protocols and Section 7 offers concluding remarks.

2. ATTACKS ON WIRELESS NETWORKS

Two kinds of attacks can be launched against ad-hoc networks [6], *passive* and *active*.

2.1 Passive Attacks

In passive attacks the attacker does not perturb the routing protocol. Instead, it only eavesdrops on the routing traffic and tries to extract valuable information like node hierarchy and network topology from it. For example, if a route to a particular node is requested more frequently than to other nodes, the attacker might expect that the node is significant for the operation of the network, and disabling it could bring down the entire network. Likewise, even when it might not be possible to isolate the exact position of a node, one may be able to find out information about the network topology by analysing the contents of routing packets. This attack is virtually impossible to detect in the wireless environment and hence also extremely difficult to prevent.

2.2 Active Attacks

In active attacks, the aggressor node has to expend some of its energy in order to carry out the attack. Nodes that perform active attacks with the aim of disrupting other nodes by causing network outage are considered to be malicious, while nodes that perform passive attacks with the aim of saving battery life for their own communications are considered to be selfish. In active attacks, malicious nodes can disrupt the correct functioning of a routing protocol by modifying routing information, by fabricating false routing information or by impersonating nodes [4].

2.2.1 Attacks Using Modification

Routing protocols for ad-hoc networks are based on the assumption that intermediate nodes do not maliciously change the protocol fields of messages passed between nodes. This assumed trust permits malicious nodes to easily generate traffic subversion and Denial-of-Service (DoS) attacks. Attacks using modification are generally targeted against the integrity of routing computations and so by modifying routing information an attacker can cause network traffic to be dropped, redirected to a different destination, or take a longer route to the destination increasing communication delays. An example is for an attacker to send fake routing packets to generate a routing loop, causing packets to pass through nodes in a cycle without getting to their actual destinations, consuming energy and bandwidth. Similarly, by sending forged routing packets to other nodes, all traffic can be diverted to the attacker or to some other node. The idea is to create a *black hole* by routing all packets to the attacker and then discarding it. As an extension to the black hole, an attacker could build a *grey hole*, in which it intentionally drops some packets but not others, for example, forwarding routing packets but not data packets. A more subtle type of modification attack is the creation of a tunnel (or *wormhole*) in the network between two colluding malicious nodes linked through a private network connection. This exploit allows a node to short-circuit the normal flow of routing messages creating a virtual vertex cut in the network that is controlled by the two colluding attackers.

2.2.2 Attacks Using Fabrication

Fabrication attacks are performed by generating false routing messages. These attacks are difficult to identify as they are received as legitimate routing packets. The *rushing attack* [8] is a typical example of malicious attacks using fabrication. This attack is carried out against on-demand routing protocols that hold back duplicate packets at every node. An attacker rapidly spreads routing messages all through the network, suppressing legitimate routing messages when nodes discard them as duplicate copies. Similarly, an attacker can nullify an operational route to a destination by fabricating routing error messages asserting that a neighbour can no longer be contacted.

2.2.3 Attacks Using Impersonation

A malicious node can initiate many attacks in a network by masquerading as another node (spoofing). Spoofing takes place when a malicious node fakes its identity by changing its MAC or IP address in order to alter the network topology outlook for a benevolent node. As an example, a spoofing attack allows the creation of loops in the routing information collected by a node, with the result of partitioning the network.

3. VULNERABILITIES IN COMMON AD-HOC NETWORK ROUTING PROTOCOLS

Ad-hoc network routing protocols usually operate at the second or third layer of the TCP/IP protocol suite. Generally speaking, a typical routing protocol works in the following manner. The source node that requires a route to a destination broadcasts a route-requesting packet. Each intermediate recipient node retransmits this packet, if it is not a duplicate. When this packet reaches the destination, it originates a route-reply packet that is unicast towards the sender. For route maintenance these protocols use route-error packets that informs the active users regarding a route failure.

The intermediate nodes, while forwarding or overhearing the routing packets, add all necessary routing information to locally maintained routing tables or cache.

The core security problems that affect ad-hoc networks, originate due to the route development by the intermediate nodes. In wired networks, usually all traffic originating of a particular subnet is routed towards a known trusted gateway. That trusted gateway usually forwards the traffic to its own trusted gateway and so on until the traffic reaches its destined subnet. However, in case of ad-hoc networks, as the nodes are mobile, there is no single traffic concentration point and no known trust entity. Therefore, in order to achieve connectivity, the neighboring and intermediate nodes have to be trusted and frequently risked for their truthful execution.

Some known security vulnerabilities that are present in common ad-hoc network routing protocols are discussed in the following sub-sections.

3.1 Dynamic Source Routing (DSR) Protocol

The Dynamic Source Routing (DSR) protocol [10] is an on-demand routing protocol. Its most interesting feature is that all data packets sent using the DSR protocol have absolutely no dependency on intermediate nodes regarding routing decisions, as each carries the complete route it traverses. When a node requires a route to a particular destination, it broadcasts a ROUTE REQUEST packet. Each recipient node that has not seen this specific ROUTE REQUEST and has no knowledge about the required destination rebroadcasts this ROUTE REQUEST after appending its own address to it. If this ROUTE REQUEST reaches the destination or an intermediate node that has a route to the destination in its ROUTE CACHE, it sends a ROUTE REPLY packet containing the complete route from the source to the destination. The source node may receive a number of such route replies and may decide to select a particular route based upon the number of hops, delay or other such criteria. All nodes forwarding or overhearing any packets must add all usable routing information from that packet to their own ROUTE CACHE. For route maintenance, intermediate nodes that find any broken route return a ROUTE ERROR packet to each node that had sent a packet over that particular route.

The DSR protocol, due to its specific implementation, is susceptible to the following attacks:

3.1.1 Routing Black Hole

In this attack an attacker sends forged routing packets so as to make other nodes route traffic to a particular node. This node now has the ability to monitor all traffic flow and to selectively modify or drop packets at own discretion.

3.1.2 Gratuitous Detour

In the Gratuitous Detour attack [6], the attacker attempts to make a route through itself appear legitimate by adding virtual nodes to the route in spite of the fact that shorter usable routes may be existent in the network.

3.1.3 Deceptive alteration of IP addresses

During propagation of the ROUTE REQUEST packet, intermediate nodes add their IP addressees to it for route creation. However, any malicious node may modify, delete or add IP addressees to create routes as per its own requirement. Doing so enables malicious nodes to launch a variety of attacks in the network including creation of wormholes and black holes.

3.1.4 Deceptive alteration of Hop Count

The hop count field of the IP packet usually informs the recipient of the total number of hops that the packet has traversed so far. Consequently, malicious nodes may increase this number so as to portray longer routes or decrease it to represent shorter routes. By doing so a malicious node is able to degrade or upgrade routes, thereby creating a topology that is most favourable to it.

3.2 Ad-hoc On-Demand Distance Vector (AODV)

Ad-hoc On-Demand Distance Vector (AODV) by Perkins et al. [16] is inherently a distance vector routing protocol that has been optimised for ad-hoc wireless networks. It is an on-demand protocol as it finds the routes only when required and is hence also reactive in nature. AODV borrows basic route establishment and maintenance mechanisms from the DSR protocol and hop-to-hop routing vectors from the DSDV protocol. To avoid the problem of routing loops, AODV makes extensive use of sequence numbers in control packets. When a source node intends communicating with a destination node whose route is not known, it broadcasts a ROUTE REQUEST packet. Each ROUTE REQUEST packet contains an ID, source and the destination node IP addresses and sequence numbers together with a hop count and control flags. The ID field uniquely identifies the ROUTE REQUEST packet; the sequence numbers inform regarding the freshness of control packets and the hop-count maintains the number of nodes between the source and the destination. Each recipient of the ROUTE REQUEST packet that has not seen the Source IP and ID pair or doesn't maintain a fresher (larger sequence number) route to the destination rebroadcasts the same packet after incrementing the hop-count. Such intermediate nodes also create and preserve a REVERSE ROUTE to the source node for a certain interval of time. When the ROUTE REQUEST packet reaches the destination node or any node that has a fresher route to the destination, a ROUTE REPLY packet is generated and unicast back to the source of the ROUTE REQUEST packet. Each ROUTE REPLY packet contains the destination sequence number, the source and the destination IP addresses, route lifetime together with a hop count and control flags. Each intermediate node that receives the ROUTE REPLY packet, increments the hop-count, establishes a FORWARD ROUTE to the source of the packet and transmits the packet on the REVERSE ROUTE. For preserving connectivity information, AODV makes use of periodic HELLO messages to detect link breakages to nodes that it considers as its immediate neighbours. In case a link break is detected for a next hop of an active route a ROUTE ERROR message is sent to its active neighbours that were using that particular route.

AODV being based on the distance vector algorithm involves maximum interaction with its immediate neighbours without any reliance on global knowledge. This limited perspective allows malicious nodes to mount a variety of modification attacks against ignorant nodes. Two of the most common attacks are listed as below:

3.2.1 Deceptive incrementing of Sequence Numbers

Destination Sequence numbers determine the freshness of a route. The destination sequence numbers maintained by different nodes are only updated when a newer control packet is received with a higher sequence number. Normally the destination sequence numbers received via control packets cannot be greater than the previous value held by the node plus one [24]. However, malicious nodes may increase this number so as to advertise fresher routes towards a particular destination. If this difference is equal or larger than two then there is a high probability that the network may be under a modification attack.

3.2.2 Deceptive decrementing of Hop Count

AODV favours route freshness over route length i.e. a node prefers a control packet with a larger destination sequence and hop count number than a control packet with a smaller destination sequence and hop count number. However, if the destination sequence numbers are the same then the route with the least hop count is given preference. Malicious nodes frequently exploit this mechanism in order to generate fallacious routes that portray minimal hop-counts.

3.3 Temporally Ordered Routing Algorithm (TORA)

Temporally Ordered Routing Algorithm by Park et al. [15] is a distributed routing protocol for multi hop networks. The unique feature of this protocol is that it endeavours to localize the spread of routing control packets. The protocol is basically an optimised hybrid of the Gafni Bertsekas [5] and Lightweight Mobile Routing [2] protocols. It guarantees loop freedom, multiple routes and minimal communication overhead even in highly dynamic environments. The protocol targets at minimizing routing discovery overhead and in doing so prefers instant routes to optimal routes. The protocol supports source-initiated on-demand routing for networks with a high rate of mobility as well as destination oriented proactive routing for networks with lesser mobility.

In the on-demand mode, TORA algorithm performs four routing functions: Route Creation, Route Maintenance, Route Erasure and Route Optimisation. To accomplish these functions it uses four distinct control packets: Query (QRY), Update (UPD), Clear (CLR) and Optimisation (OPT). During route discovery, a source node requiring a route to a destination, broadcasts a QRY packet containing the destination address. The QRY packet is propagated through the network until it reaches the destination or any intermediate node possessing a route to the intended destination. The recipient of the QRY packet broadcasts an UPD packet that lists its height with respect to the destination. If the destination itself replies to a QRY packet it sets the height to zero in the UPD packet. Each node that receives the UPD packet sets its own height greater than that in the UPD packet. This results in creation of a directed acyclic graph (DAG) with all links pointing in the direction of the destination as the root. In the proactive mode, routes are created using the OPT packet that is sent out by the destination. The OPT packet, which is similar to the UPD packet, also consists of a sequence number for duplication avoidance. Each recipient nodes adjusts its height data structure and sends out a OPT packet to neighbouring nodes.

Whenever a partition is spotted, the node sets its own and the height of all its neighbours to NULL and broadcasts a CLR packet. The neighbouring nodes that receive the CLR packet, based upon the reference level, also set the heights in a similar manner and rebroadcast the CLR packet. In this way the height of each node in the portion of the network that is partitioned is set to NULL and all invalid routes are erased.

As each node in TORA maintains multiple DAG's to the destination so in any network with an average n number of nodes each with n/2 downstream neighbours, a node could still effectively communicate with the destination node upon link failure of (n/2)–1 nodes. However, to sustain this redundancy, each node maintains a height data structure, link status along with a number of state and auxiliary variables for each destination node.

TORA is not a standalone routing protocol but requires the services of Internet MANET Encapsulation Protocol (IMEP). IMEP [3] has been designed as a network layer protocol that provides link status, neighbour connectivity information, address resolution and other services to Upper Layer Protocols (ULP).

Some of the common attacks specific to the TORA protocol are:

3.3.1 Deceptive alteration of QRY and UPD Packets

During propagation of the QRY and UPD packets, any intermediate nodes may modify the destination IP address or the height quintuple so as to create fallacious directed acyclic graphs. Doing so enables malicious nodes to launch a variety of attacks in the network including creation of wormholes and black holes.

3.3.2 Malicious propagation of CLR packets

Malicious nodes may propagate illegitimate CLR packets so as to erase genuine routes. By doing so a malicious node is able to render inoperative certain critical routes and launch other attacks against the network.

4. SECURITY ISSUES

Some of the issues that need to be addressed before implementing any security scheme in ad-hoc networks are:

1. Energy is one of the greatest constraints to a node's capabilities.
2. Symmetric encryption/decryption algorithms and hashing functions consume minimal computational energy in comparison to public key algorithms.
3. Transmission energy consumption is over three orders of magnitude greater than the energy consumption for encryption and hashing [1].
4. It is problematical to establish and sustain a trusted third party in the ad-hoc networks.

5. Intra-node relationships are usually less formal, temporary and short-term.
6. There are two types of adversaries that an ad-hoc network may have to deal with: malicious and compromised.

Nodes in a wireless ad-hoc network are extremely dependent upon efficient utilization of their battery packs. Undue usage due to extra transmissions or computing can result in rapid battery draining. A less vivid but understated malicious behaviour is node selfishness in which nodes, in order to save their batteries, may be tempted to not relay packets. An easy solution against such attacks is the establishment of a trusted third party that can facilitate building of trust relationships among communicating nodes. Public Key Infrastructure (PKI) is an effective way of establishing trust but is deemed unsuitable because it makes use of asymmetric cryptographic algorithms that have been known to be a target of Denial of Service attacks. Also, maintaining a centralized or distributed repository of certificates is itself a potential target of attack.

In a mobile ad-hoc network as the nodes are constantly leaving and entering the network the relationships between the nodes tend to be short lived and reciprocal in nature. As the time span of such relationships is really acute, a benevolent behaviour is expected of all nodes. However, in addition to benevolent nodes, there are also malicious and compromised nodes present in the network. Malicious nodes attempt to eavesdrop, replay, distort and impersonate messages while a compromised node is a benevolent node that has been taken over by an adversary. Compromised nodes produce valid signatures, identification and certificates and are hence very difficult to isolate. Intrusion detection systems such as those proposed by Zhang et al. [27] and Kachirski et al. [11] can be employed for identifying such nodes in ad-hoc networks.

5. SECURE ROUTING PROTOCOLS

Many Mobile Ad-hoc Network (MANET) protocols have been designed without having security in mind, based upon the assumption that all the nodes in the network are friendly. The actual scenario is quite different from this assumption as in a MANET a hostile node may join the network at any time as a friend, or any friendly node may become hostile. This malicious or hostile node can disrupt the network routing services and, in the worst-case scenario cause, a complete failure of the network communications.

The provisioning of security services to MANETs face a totally different set of problems. These problems include the lack of security in the medium, energy and processing constraints and poor physical protection of the nodes.

Nonetheless, the absence of a fixed infrastructure in MANET context ensures that no part of the network is committed to support independently any precise network functionality like routing, neighbour discovery, etc. Additionally the lack of centralized services like Key Distribution Centres (KDC), Certification Authorities (CA) and Name Servers complicate the introduction of security in the perspective of ad-hoc networks.

Routing in MANETs encompasses two major problems. One problem is to ensure that data is routed securely through trusted nodes and the second is the security of the routing protocol messages. In view of the fact that both data and control messages utilize the same wireless transmission medium, routing protocol messages can be modified to alter routing behaviour. This means that if a routing protocol message is altered to generate a false route, then no amount of security on data packets can correct this routing misbehaviour. This means that the security of the routing protocols is imperative for secure routes through the ad-hoc network. A good secure routing protocol must conform to the following requirements [4] to ensure that the discovered path from source to destination functions properly in the presence of malicious nodes:

- Authorized nodes to perform route computation and discovery
- Minimal exposure of network topology
- Detection of spoofed routing messages
- Detection of fabricated routing messages
- Detection of altered routing messages
- Avoiding formation of routing loops
- Prevent redirection of routes from shortest paths

To achieve the said goals in ad-hoc networks, secure routing protocols use a combination of Hash functions and Asymmetric or Symmetric cryptographic algorithms [18]. Hash functions, also known as one-way functions, take in a variable length input called the pre-image and convert it into a smaller fixed-length output called the Hash. The security of this output hash lies in the one-way property of the hash function, which makes it computationally impossible to find the pre-image from the hash in the reverse direction. A hash function is said to be cryptographically strong if it is computationally impossible to find another pre-image for the same hash value, also called a collision. Hash functions are extremely efficient and are generally employed in ad-hoc networks for integrity verification and authentication of data and control packets.

Asymmetric algorithms, also allow nodes to encrypt network traffic using public keys so that the data is only recoverable by the possessor of the corresponding private key. The private key is kept secure by every node and the public keys are distributed using trusted third party servers or mutual sharing techniques [28]. Cryptosystems based on asymmetric algorithms use a variety of NP hard problems including composite number factorisation and discrete log problem [12], making them very secure and reliable. However, the actual implementation of such algorithms incurs colossal computation requirements and is generally deemed infeasible for ad-hoc network environments.

In contrast, Symmetric algorithms make use of a single key for encryption/decryption and are relatively a thousand times faster than Asymmetric cryptographic algorithms. Hash functions, when used in conjunction with symmetric cryptographic algorithms prove extremely beneficial for securing the routing process and data in ad-hoc networks. Moreover, as the hashing process is extremely efficient and secure, it is used to counter a variety of attacks launched against the availability of nodes in an ad-hoc network.

The following sub-sections discuss some contemporary secure routing protocols that use cryptographic algorithms to protect an ad-hoc network:

5.1 ARAN

The Authenticated Routing for Ad-hoc Networks (ARAN) secure routing protocol [4] is an on-demand routing protocol that identifies and shields against malevolent actions by malicious nodes in the ad-hoc network environment. ARAN relies on the use of digital certificates and can successfully operate in the managed-open scenario where no network infrastructure is pre-deployed, but a small amount of prior security coordination is expected. ARAN provides authentication, message integrity and non-repudiation in ad-hoc networks by using a preliminary certification process that is followed by a route instantiation process that guarantees end-to-end provisioning of security services.

ARAN requires the use of a trusted certificate server (T). All nodes are supposed to keep fresh certificates with the trusted server and should know T's public key. Prior to entering the ad-hoc network, each node has to apply for a certificate that is signed by T.

$$T \rightarrow A : cert_A = [IP_A, K_{A+}, t, e]K_{T-}$$

The certificate contains the IP address of the node, its public key, a timestamp of when the certificate was generated and a time at which the certificate expires, along with the signature by T.

ARAN accomplishes the discovery of routes by a broadcast route discovery message from a source node, which is replied to in a unicast manner by the destination node. All the routing messages are authenticated at every hop from the source to the destination, as well as on the reverse path from destination to source. Each node along the path performs the following steps: -

- Validates the previous node's signature
- Removes the previous node's certificate and signature
- Records the previous node's IP address
- Signs the original contents of the message
- Appends its own certificate
- Forward broadcasts the message

ARAN also provides similar measures for authenticated route set-up, route maintenance and key revocation through the use of certificates.

5.2 ARIADNE

ARIADNE [6] is an on-demand secure ad-hoc routing protocol based on the Dynamic Source Routing (DSR) protocol that protects against node compromise and relies only on extremely efficient symmetric cryptography. The security of ARIADNE is based upon the secrecy and authenticity of keys that are kept at the nodes. ARIADNE prevents a large number of Denial-of-Service attacks from malicious or compromised nodes.

ARIADNE provides assurance that the target node of a route discovery process can verify the initiator, that the initiator can verify each transitional node that is on the path to the destination present in the ROUTE REPLY message and that no intermediate node can reduce the node list in the ROUTE REQUEST or ROUTE REPLY messages. Route Discovery is performed in two stages: the Initiator floods the network with a ROUTE REQUEST that solicits a ROUTE REPLY from the Target. During route discovery the Target authenticates each node in the node list of the ROUTE REQUEST and the Initiator authenticates each individual node in the node list of the ROUTE REPLY. For node authentication, ARIADNE has three alternative techniques i.e. TESLA (Timed Efficient Stream Loss tolerant Authentication), Digital Signatures, or pair-wise shared secret keys.

TESLA [17] is a broadcast authentication protocol for authenticating messages. It is different from traditional asymmetric protocols as it achieves its asymmetry from clock synchronization and delayed key disclosure. When used for authentication, each sender chooses a random key and generates a one-way key chain by repeatedly computing a one-way hash function on the initial key. The sender publishes each key of the key chain at a pre-determined key disclosure interval in the reverse order from its generation. The published key remains secret for the next key disclosure delay interval. The sender computes the TESLA key using a pseudo-random function on the hash key chain. This key is used to compute the Message Authentication Code (MAC) for any packet to be transmitted. When the receiver receives a packet, it buffers it until the sender discloses the required element of the key chain. To authenticate any received value on the one-way chain, a node verifies it by hashing it a number of times to determine if the computed value matches a previous known authentic key on the chain. If the key chain element is authentic then the recipient computes the corresponding TESLA key using the pseudo-random function on the hash key chain, calculates the MAC and compares the result with the MAC in the received packet. If the MAC's match, it ensures the authenticity of the packet and the sender.

When TESLA is used for authentication, each sender picks a random initial key K_n and creates a one-way hash chain by repeatedly applying a hash function H on the preliminary value:

$$K_{n-1} = H[K_n],\ K_{n-2} = H[K_{n-1}],\ \ldots.$$

In general,

$$K_i = H[K_{i+1}] = H^{n-i}[K_n]$$

The one-way hash chain provides the property that another node can only increase a metric in a routing update but cannot decrease it. To verify any obtained value of the one-way chain, a node computes any preceding key K_i from a key K_j, where $i > j$, using the following equation to determine if the computed value equals a previously known authentic key on the chain.

$$K_j = H^{i-j}\,[K_i]$$

Each key of the key chain is also published by the sender at known intervals in the reverse order from its creation i.e. $K_0, K_1, \ldots, K_n$.

With TESLA, each hop authenticates the information in the ROUTE REQUEST packet during route discovery. The ROUTE REPLY is buffered at the target until intermediate nodes release the corresponding TESLA keys. When the TESLA security condition is met at the target, a Message Authentication Code (MAC) is included in the ROUTE REPLY. When the initiator receives the ROUTE REPLY it verifies that the target MAC is valid including certain other verifications. If all the tests are successful the initiating node accepts the ROUTE REPLY. When route discovery is performed using digital signatures, the MAC list in the ROUTE REQUEST becomes a signature list, where the data that was used to compute the MAC is instead used to compute a signature. Route Discovery using MACs is most efficient but it requires that all nodes possess a pair of shared keys. The MAC list in the ROUTE REQUEST is computed by means of a key that is shared between the target and the current node. This MAC is authenticated at the target and not returned in the ROUTE REPLY. Similarly, during route maintenance, to prevent unauthorized nodes from sending ROUTE ERROR packets, ARIADNE requires that the sender authenticate a ROUTE ERROR. If this authentication is delayed then each node that is able to authenticate the ROUTE ERROR packet, buffers it until the corresponding keys are received.

5.3 SAODV

The Secure Ad-hoc On-Demand Distance Vector (SAODV) by Zapata [26] is an extension of the AODV routing protocol. It can be used to protect the route discovery mechanism of AODV by providing security features like integrity, authentication and non-repudiation. The protocol operates mainly by using new extension messages with the AODV protocol. In these extension messages there is a signature produced by digesting the AODV packet using the private key of the original sender of the Routing message.

The Secure-AODV scheme is based on the assumption that each node possesses certified public keys of all network nodes. Ownership of certified public keys enables intermediate nodes to authenticate all in-transit routing packets. The originator of a routing control packet appends its RSA signature and the last element of a hash chain to the routing packets. As the packets traverse the network, intermediate nodes cryptographically authenticate the signature and the hash value. The intermediate nodes generate the kth element of the hash chain, with k being the number of traversed hops, and place it in the packet. The route replies are supplied either by the destination or intermediate nodes having an active route to the required destination.

The SAODV protocol gives two alternatives for ROUTE REQUEST and ROUTE REPLY messages. In the first case when a ROUTE REQUEST is sent, the sender creates a signature (i.e. encryption using the private key of the sender of all the fields in the AODV packet less the hop count) and appends it to the packet. Intermediate nodes authenticate the signature before creating or updating the reverse route to that host. The reverse route is stored only if the signature is verified. When this packet reaches the final destination, the node signs the ROUTE REPLY with its private key and sends it back. The intermediate and final nodes, again verify the signature before creating or updating a route to that host. The signature of the sender is also stored along with the route entry. The second case is also similar to the first one with the only disparity being that the ROUTE REQUEST message has another signature that is always stored along with the reverse route. This second signature is used in the regular and gratuitous ROUTE REPLYs to future ROUTE REQUESTs that the node might reply to as an intermediate node.

5.4 SAR

Security-Aware Ad-hoc Routing (SAR) by Yi et al. [25] is a generalized framework for any on-demand secure ad-hoc routing protocol. SAR uses security information to dynamically control the choice of routes installed in the routing tables. SAR enables applications to selectively implement a subset of security services based on a cost-benefit analysis.

SAR requires that nodes having the same trust level must share a secret key. SAR augments the routing process using hash digests and symmetric encryption mechanisms. The signed hash digests provide message integrity while the encryption of packets ensures their confidentiality. The routes discovered by SAR may not always be the shortest between any two communicating entities in terms of hop-count. However, these routes do have a quantifiable guarantee of security. If there is more than one route that satisfies the required security attributes, then SAR will find the shortest such route. SAR will find the optimal routes if all the nodes on the shortest path can satisfy the security requirements. However, if the ad-hoc network does not have a path with nodes that meet the security requirements, SAR will fail to find a route even if the network is connected. SAR, when implemented on the AODV protocol, adds two additional fields to the ROUTE REQUEST packet and one additional field to the ROUTE REPLY packet. The first field is the security requirement (RQ SEC REQUIREMENT) and is set by the sender. It indicates the preferred level of trust for the path to the destination. This field can be used to carry simple integer values or bit vectors to reflect the existing hierarchies or combinations of different types of security services. For example, a three-bit vector can be used to specify if the nodes desire to do a simple hash, digital signature, or content encryption over the AODV routing packets. The second field added to the ROUTE REQUEST packet is the security guarantee (RQ SEC GUARANTEE) that signifies the maximum level of security provided by the discovered paths. RQ SEC GUARANTEE is updated at every hop during the route discovery phase. If RQ SEC REQUIREMENT has an integer representation then the RQ SEC GUARANTEE will be the minimum of all the security levels of the participating nodes in the path. However, if RQ SEC REQUIREMENT is represented in bit vectors, RQ SEC GUARANTEE will be computed by ANDing the RQ SEC REQUIREMENT values of the participating nodes in the path. The value thus computed is copied into the additional RP SEC GUARANTEE field of the ROUTE REPLY packet and sent back to the sender. The sender can use this value to establish the security level over the entire path. This value indicates the actual level of trust the path provides so the sender can either use this security guarantee value or determine whether a more secure connection is required. This value is also copied into the routing tables of the nodes in the reverse path, to preserve security information with reference to cached paths.

5.5 SEAD

The Secure Efficient Distance vector (SEAD) by Hu et al. [7] protocol is a proactive secure routing protocol based on the Destination-Sequenced Distance Vector protocol (DSDV). SEAD is specifically based on the DSDV-SQ (Sequence Number) version of the DSDV protocol. The DSDV-SQ version of the DSDV protocol prevents routing loops that are caused by out of order updates. Routing can be disrupted if a malicious node modifies the sequence number or the metric field of a routing table update message. SEAD deals with such attacks that modify routing information during the update phase of the DSDV-SQ protocol. It also provides measures to thwart Replay attacks.

SEAD is basically designed to authenticate the sequence number and metric of a routing table update message using hash chains elements. In addition, the receiver of SEAD routing information also verifies the sender, making sure that the routing information originates from the right node. Each node uses a specific authentic element that is signed from its hash chain in each routing update that it sends about itself. This initial element of the one-way hash chain offers authentication for the lower bound on the metric in other routing updates for that node. The utilization of a hash value, analogous to the sequence number and metric in a routing update entry, foils any node from promoting a route to some destination with a greater sequence. Similarly, a node cannot present a route better than those for which it has already received an advertisement because the metric in an existing route cannot be decremented owing to the one-way nature of the hash chain. When a routing update is received by a node, it verifies the authenticity of the information for each entry in the update by means of the destination address, the sequence number and the metric of the received entry, along with the most recent prior authentic hash value received from that destination's hash chain. The received elements are hashed the correct number of times to verify the authenticity of the received information. If the computed hash value and the real hash value match, the entry is considered to be authentic and the node processes it in the routing algorithm as a standard received routing update entry; if not, the node disregards the entry and does not adjust its routing table.

5.6 SLSP

Secure Link State Routing Protocol (SLSP) by Papadimitratos et al. [14] provides secure proactive topology discovery and can be used either as a stand-alone protocol or as part of a hybrid routing framework when combined with a reactive protocol. It can operate in networks of recurrently changing topology and memberships. It is resilient against individual attackers and it is capable of altering its range between local and network-wide topology discovery.

To function effectively without a central key management authority, SLSP enables each node to periodically broadcast its public key to nodes within its zone. In addition each node also broadcasts signed HELLO messages containing its Medium Access Control (MAC) address and IP address (MAC, IP) pair to its neighbours. The distribution of MAC addresses strengthens the scheme by forbidding nodes from spoofing at the data link layer. It also assists in protection against flooding DoS attacks. To achieve these goals, a Neighbour Lookup Protocol (NLP) is made an integral part of SLSP. NLP is responsible for the following tasks:

- Maintaining a mapping of MAC and IP layer addresses of the node's neighbours.
- Identifying potential discrepancies, such as the use of multiple IP addresses by a single data-link interface.
- Measuring the rates at which control packets are received from each neighbour by differentiating the traffic primarily based on MAC addresses.

This rate of incoming control packets helps in discarding nodes which maliciously seek to exhaust network resources. SLSP requires that the MAC layer 48-bit address be passed to the network layer. This requires a straightforward alteration of the device driver, so that the data link address is "passed up" to the routing protocol with every packet. NLP then extracts and maintains the 48-bit hardware source address for each received (overheard) frame, together with the encapsulated IP address. The mappings amid these two addresses are kept in the table until transmissions from the corresponding neighbouring nodes are overheard or the lost neighbour timeout period expires.

Link State updates (LSU) are recognized by the IP address of their originator and a 32-bit sequence number. To ensure that the LSUs only propagate in a zone of its origin, receiving nodes verify if they have the public key of the originating node, except if the key is provided with the LSU. The LSU is then verified, its hop chain updated, TTL decremented and rebroadcasted. An LSU is discarded based upon NLP notification or an error in the hop hash chain. A hash chaining mechanism is used to ensure that the hop counters are authenticated.

SLSP employs a round robin servicing mechanism to provide the assurance that benign control traffic will be relayed even under clogging DoS attacks. To realize this mechanism, nodes maintain a priority standing of their neighbours according to the rate of control traffic experimented by

NLP. The top priority is assigned to the nodes that are generating or relaying link state updates with the lowest pace and vice versa.

5.7 SRP

The Secure Routing Protocol (SRP) by Papadimitratos et al. [13] is an extension for reactive routing protocols. The scheme is robust in the presence of multiple non-colluding nodes, and provides precise routing information in a timely manner. SRP counters attacks that disrupt the route discovery mechanism and warrants the acquisition of accurate topological information. By using SRP, a node initiating a route discovery is able either to identify and discard replies providing fake routing information, or avoids getting them. The fundamental assumption is the presence of a security association (SA) between the source node (S) and the destination node (T). The trust relationship could be established by possession of the public key of the other communicating end and then agreeing upon a shared secret key ($K_{S,T}$) between each other.

SRP based on the dynamic source routing protocol (DSR) necessitates the addition of a 6-word header that contains unique identifiers that tag the discovery process and a Message Authentication Code (MAC). While initiating a ROUTE REQUEST the source node has to compute a MAC using a keyed hash algorithm that accepts as input the entire IP header, the ROUTE REQUEST packet and the shared key $K_{S,T}$. The transitional nodes that pass on the ROUTE REQUEST towards the destination calculate the frequencies of queries received from their neighbours in order to control the query propagation process. Each node maintains a priority ranking that is inversely proportional to the queries' rate in order to realize flow control. Any node overwhelming the network with unsolicited ROUTE REQUESTs will be served last due to the low priority ranking mechanism. When a ROUTE REQUEST is received, the destination node confirms the integrity and authenticity of the ROUTE REQUEST by computing the keyed hash of the required fields and then comparing it with the MAC contained in the SRP header. If the ROUTE REQUEST is legitimate, the destination responds with a ROUTE REPLY using the SRP header in a manner similar to the ROUTE REQUEST. The source node verifies the query identifiers and MAC integrity values of all ROUTE REPLY packets in order to avert any replay attacks.

6. COMPARISON

A comparison of the current secure ad-hoc routing protocols is shown in the Table 4-1. It reveals that most of the protocols have been configured to operate with reactive ad-hoc routing protocols. These protocols make use of symmetric and asymmetric cryptography to achieve the different security parameters.

Table 4-1. Comparison of Secure Routing Protocols for Ad-hoc Networks

Performance Parameters	ARAN	ARIADNE	SAODV	SAR	SEAD	SLSP	SRP
Type	Reactive	Reactive	Reactive	Reactive	Proactive	Proactive	Reactive
MANET Protocol	AODV/ DSR	DSR	AODV	AODV	DSDV	ZHLS	DSR/ ZRP
Encryption	Asym	Sym	Asym	Sym/ Asym	Sym	Asym	Sym
Synchronization	No	Yes	No	No	Yes	No	No
Trust Authority	CA	KDC	CA	CA/ KDC	CA	CA/ KDC	CA
Authentication	Yes	Yes	Yes	Yes	Yes	Yes	Yes
Confidentiality	Yes	No	No	Yes	No	No	No
Integrity	Yes	Yes	Yes	Yes	No	No	Yes
Non-repudiation	Yes	No	Yes	Yes	No	Yes	No
Anti - Spoofing	Yes	Yes	Yes	Yes	No	Yes	Yes
DoS Attacks	No	Yes	No	No	Yes	Yes	Yes

Protocols using public keys provide a range of security services but may be computationally expensive to operate in real environments. This is due to the fact that the use of public-key cryptography imposes a high processing overhead on the intermediate nodes and can be considered unrealistic for a wide range of network instances. An inefficient authentication system could become a target of a Denial-of-Service (DoS) attack by an attacker that floods nodes with malicious authentication messages so as to saturate them with computationally expensive modulo operations. The prevention of DoS attacks in a wireless network is quite a difficult and challenging task. Some protocols have tried to avoid this attack by using symmetric algorithms and priority mechanisms to ensure availability of nodes.

Most of the secure protocols address attacks launched by a number of non-colluding malicious nodes. However, attacks by multiple colluding nodes, like wormhole or vertex cut, are difficult to detect and solutions to these attacks have only been suggested by ARIADNE.

For node authentication ARIADNE and SEAD suggest using the TESLA broadcast authentication scheme with delayed key disclosure. However, as TESLA requires a clock synchronization mechanism, which in turn is also prone to attacks, it is considered to be an unrealistic requirement for ad-hoc networks.

For securing the mutable information in routing messages, ARIADNE, SAODV, SEAD and SLSP make use of hash chains, which is an efficient way to obtain authentication.

In hostile environments, the network topology must not be exposed to adversaries by the routing messages. Exposure of the network topology may be an advantage for adversaries trying to destroy or capture nodes. The confidentiality of routing messages has been taken into account by the ARAN and SAR routing protocols.

A common observation regarding all protocols is the assumption of a central trust authority for key management or the existence of pre-shared keys. This assumption has been made due to the absence of a central authorization facility in an open and distributed communication environment, which requires a cooperative operation. To justify the creation of such an entity in an improvised environment the term "managed ad-hoc network" has been introduced where such an entity can be established or the nodes can be pre-configured with encryption keys before joining a network.

Most of the current key exchange protocols use the central trust authority for initial authentication. A variant of the central trust authority is the Distributed Public-Key Model [28] that makes use of threshold cryptography to distribute the private key of the Certification Authority (CA) over a number of servers. Whatever the case may be, the requirement of a trust authority in such a dynamic environment is considered both impractical and unsafe. Since, such an entity may not be always accessible and it also creates a single point of failure. Similarly, key exchange using a Key Distribution Server [20] creates a similar set of problems.

Employing cryptographic mechanisms is another way of stimulating trust into the network, where the trust is being placed in the trusted third party and the encryption algorithm. However, as the nodes in an ad-hoc network are mobile in an open environment, they are more vulnerable to capture due to limited physical protection. Consequently, the compromise of a key repository or an authentication server can jeopardize the complete security infrastructure. The requirement of an omnipresent, and often omniscient, trust authority does resolve many of the core security issues in wired networks but it is neither practical nor feasible in an ad-hoc network.

Some of the authors have discussed the provisioning of specific security services by their protocols while others have not clearly defined the security coverage of their proposed protocols. As a result, a number of security services are claimed as being provided by most of these protocols. For example, some protocols only provide selective-field confidentiality while others provide connectionless confidentiality. The same may be reported for authentication, non-repudiation and integrity security services. As there is no universal agreement about many of the terms used in security literature [23], so for comparison purposes we have used security services as defined by ITU X.800 [9] and RFC 2828 [22].

The security protocols presented in this chapter are a practical response to specific problems that rise due to attacks on ad-hoc network routing protocols. Consequently, the proposed solutions only cover a subset of all possible threats and are not flexible enough to be integrated with each other.

7. CONCLUSION

The area of ad-hoc networking has been receiving increasing attention among researchers in recent years, as the available wireless networking and mobile computing hardware bases are now capable of supporting the promise of this technology. Specially configured routing protocols have been developed for these networks keeping in mind both performance and endurance. The correct execution of these routing protocols is mandatory for smooth functioning of an ad-hoc network. Similarly, the data traversing the wireless medium needs to be protected against passive eaves dropping. Over the past few years, a variety of new routing protocols targeted specifically at the security aspect of ad-hoc networking environment have been proposed, but little performance information on each protocol and no performance comparison between these protocols has previously been available.

In this chapter we have provided the description of several secure ad-hoc network routing protocols. We have presented a comparison of these protocols by highlighting their features, differences and characteristics. While it is not clear that any particular algorithm or class of algorithm is best for all scenarios, each protocol has definite advantages and disadvantages, and is appropriate for application in a number of secure ad-hoc network environments.

REFERENCES

1. Carman, D. W., Kruus, P. S., and Matt B. J., "Constraints and approaches for distributed sensor network security," *Technical Report #00-010*, NAI Lab. 2000,
2. Corson, M. S., and Ephremides, A., "Lightweight mobile routing protocol (LMR) : A distributed routing algorithm for mobile wireless networks," *Wireless Networks*, 1(1):61-81,1995.
3. Corson, S., Papademetriou, S., Papadopoulos, P., Park, V., and Qayyum, A., "An Internet MANET encapsulation protocol (IMEP) specification," *IETF MANET, Internet Draft* (work in progress), draft-ietf-manet-imep-spec02.txt., 1999.
4. Dahill, B., Levine, B. N., Royer, E., and Shields, C., "ARAN: A secure routing protocol for ad-hoc networks," *Proc. of 10th IEEE International Conference on Network Protocols (ICNP'02)*, pp. 78- 87, 2002.
5. Gafni, E., and Bertsekas, D., "Distributed algorithms for generating loop-free routes in networks with frequently changing topology," *IEEE Transactions on Communications*, 29(1):11-18, 1981.
6. Hu, Y-C., Perrig, A., and Johnson, D. B., "Ariadne: A secure on-demand routing protocol for ad-hoc networks," *Proc. of 8th ACM International Conference on Mobile Computing and Networking (MobiCom'02)*, pp. 12-23, 2002.

7. Hu, Y-C., Perrig, A., and Johnson, D. B., "SEAD: Secure efficient distance vector routing for mobile wireless ad-hoc networks," *Proc. of Fourth IEEE Workshop on Mobile Computing Systems and Applications (WMCSA'02)*, pp. 3-13, 2002.
8. Hu, Y-C., Perrig, A., and Johnson, D. B., "Rushing attacks and defense in wireless ad hoc network routing protocols," *Proc. of the ACM Workshop on Wireless Security(WiSe'03)*, pp. 30-40, 2003.
9. ITU, X.800 : *Security architecture for open systems interconnection for CCITT applications*, 1991.
10. Johnson, D. B., Maltz, D. A., and Hu, Y., "The dynamic source routing protocol for mobile ad-hoc networks (DSR)," *IETF MANET, Internet Draft* (work in progress), draft-ietf-manet-dsr-09.txt, 2003.
11. Kachirski, O., and Guha, R., "Intrusion detection using mobile agents in wireless ad-hoc networks," *IEEE Workshop on Knowledge Media Networking (KMN'02)*, pp. 153-158, 2002.
12. Menezes, A., VanOorschot, P., and Vanstone, S., *Handbook of Applied Cryptography*, CRC Press, 1996.
13. Papadimitratos, P. and Haas, Z., "Secure Routing for Mobile Ad Hoc Networks," *Proc. of Communication Networks and Distributed Systems Modelling and Simulation Conference (CNDS'02)*, pp. 1-13, 2002.
14. Papadimitratos, P., and Haas, Z., "Secure link state routing for mobile ad-hoc networks," *Proc. of Symposium on Applications and the Internet Workshops (SAINT'03)*, pp. 379-383, 2003.
15. Park, V., and Corson, M., "Temporally-ordered routing algorithm (TORA) version 1 functional specification," *IETF MANET, Internet Draft* (work in progress), draft-ietf-manet-tora-spec-04.txt, 2001.
16. Perkins, C., Royer, E. B., and Das, S., "Ad-hoc on-demand distance vector (AODV) routing," *IETF MANET, RFC 3561*, 2003.
17. Perrig, A., Canetti, R., Tygar, D., and Song, D., "The TESLA broadcast authentication protocol," *RSA CryptoBytes*, pp. 2-13, 2002.
18. Pirzada, A. A., and McDonald, C., "A review of secure routing protocols for ad-hoc mobile wireless networks", *Proc. of 7th International Symposium on DSP for Communication Systems (DSPCS'03)* and 2nd *Workshop on the Internet, Telecommunications and Signal Processing (WITSP'03)*, pp. 118-123, 2003.
19. Pirzada, A. A., and McDonald, C., "Establishing trust in pure ad-hoc networks," *Proc. of 27th Australasian Computer Science Conference (ACSC'04)*, 26(1): 47-54, 2004.
20. Pirzada, A. A., and McDonald, C., "Kerberos assisted authentication in mobile ad-hoc networks," *Proc. of 27th Australasian Computer Science Conference (ACSC'04)*, 26(1): 41-46, 2004.
21. Royer, E. M., and Toh, C-K, "A review of current routing protocols for ad-hoc mobile wireless networks," *IEEE Personal Communications Magazine*, 6(2), pp. 46-55, 1999.
22. Shirey, R., " Internet Security Glossary," *IETF RFC 2828*, 2000.
23. Stallings, W., *Network Security Essentials 2nd Edition*, Prentice Hall, 2003.
24. Wang, W., Lu, Y., and Bhargava, B., "On vulnerability and protection of ad-hoc on-demand distance vector protocol," *Proc. of International Conference on Telecommunication (ICT'03)*, 1:375-382, 2003.
25. Yi, S., Naldurg, P., Kravets, R., "Security aware ad-hoc routing for wireless networks," *Proc. of the 2nd ACM International Symposium on Mobile ad hoc networking and computing (MobiHoc'01)*, pp. 299-302, 2001.
26. Zapata, M. G., "Secure ad-hoc on-demand distance vector (SAODV) routing," *IETF MANET, Internet Draft* (work in progress), draft-guerrero-manet-saodv-00.txt, 2001.

27. Zhang, Y. and Lee, W., "Intrusion detection in wireless ad-hoc networks," *Proc. of the 6th ACM International Conference on Mobile Computing and Networking, (MobiCom'00)*, pp. 275-283, 2000.
28. Zhou, L. and Haas, Z. J., "Securing ad-hoc networks," *IEEE Network Magazine*, 13(6):24-30, 1999.

Chapter 5

CROSS LAYER DESIGN FOR AD-HOC NETWORKS

Peter Pham, Sylvie Perreau and Aruna Jayasuriya
Institute for Telecommunications Research
University of South Australia, SA5095, Australia

Abstract: Research on ad hoc networks has recently attracted a lot of interest in the communication community. However, research on the performance of ad hoc networks under Rayleigh fading channel is still in its infancy. Some works have indicated that the performance of ad hoc networks could be badly affected by Rayleigh fading channel [1]. In this paper, we identify the causes for this performance degradation and we propose a new cross-layer approach which improves the network throughput, decreases unnecessary packet transmissions, saves power and bandwidth resource, and reduces packet loss due to channel contention under IEEE 802.11 DCF. In addition, we also propose two new Markovian models: one characterising Rayleigh fading channels and the other one modelling the IEEE 802.11 DCF. Both models are then combined to study the performance of the ad hoc networks MAC layer under Rayleigh fading channel conditions. The performance of the new cross-layer approach is studied through simulations as well as through analytical models. The simulation results reveal an increase in network throughput, decrease in unnecessary packet retransmission and reduction in packet loss due to network contention. The analytical results show a close match with the simulation results.

Key words: Ad Hoc Networks, Rayleigh Fading Channel, Cross Layer Design, IEEE 802.11 DCF, Markov

1. INTRODUCTION

Recently, cross-layer design for wireless networks and ad hoc networks has attracted a lot of interest. The principle of cross-layer design is to share information across various protocol layers. Various approaches for this have been proposed [2--5]. In [3], "path-coupling" degree can be used as a criterion for route selection. This method is shown to increase the network throughput and reduce the energy consumption. In [4], it is shown that Medium Access Control (MAC) performance is strongly dependent on the choice of the accompanying routing protocol. In [5], a simple spectrally efficient rate adaptation scheme is proposed which enhances the accuracy of routing decisions.

In this paper, we propose a method based on cross layer design, that makes use of the predictability of slow Rayleigh fading channels to improve the network performance. It is important to stress out the fact that ad hoc networks based on the 802.11 Physical layer specifications can support in theory data rates of the order of several Megabits per second. Therefore, such channels would tend to be frequency selective [6]. However, in this paper, we concentrate our study on flat fading channels and we justify this assumption by the following arguments: firstly, experimental studies show that it is not useful to transmit data at rates higher than 400kbits per second as the resulting overall network throughput does not increase due to both physical layer issues (inter symbol interference) and MAC layer limitations (collisions) [7]. Therefore, if we only transmit at 400kbits per second, this would correspond to a flat fading channel.

Secondly, although studies have shown that higher data rates are achievable in ad hoc networks if Orthogonal Frequency Division Multiplexing (OFDM) or Code Division Multiplexing Access (CDMA) are used, one can easily show that with these techniqies the resulting channel is flat fading. For OFDM, the overall frequency selective channel is split into several flat fading channels. As for CDMA, the issue of multipath is solved using the RAKE receiver which again allows to deal with a flat fading channel after the despreading operation. Therefore, we can justify the assumption of flat fading used throughout this paper. Note that we still assume the 802.11 MAC layer specifications. Using this Rayleigh fading assumption, we will see that it is possible to predict when the channel is about to undergo a fade, i.e. when the receiver is unable to receive data packets correctly due to low received power. The physical layer feeds this information back to the upper layers. The upper layers can then suspend the transmission until the channel gets out of the fade, i.e. when the packet is received correctly.

In this paper, we make use of the predictability of a slow Rayleigh fading channels to improve the network performance. Specifically, it is possible to predict when the channel is in a fade, i.e. when the receiver is unable to receive data packets correctly due to low received power. The physical layer feeds this information back to the upper layers. The upper layers can then suspend the transmission until the channel gets out of the fade, i.e. when the packet is received correctly.

The paper is organized as follows. In Section 2, we present in details the innovative cross-layer approach. The simulation results of the new approach are presented in Section 3. In Section 4, a new Markov model for a Rayleigh channel is proposed. We then propose a new comprehensive mathematical model to analyse IEEE 802.11 DCF in Section 5. We also use the results from Section 4 to analyse how our cross-layer design improves the performance of IEEE 802.11 DCF. They are validated against the analytical results in Section 6. Finally, we summarize our achievements and propose future research directions in Section 7.

2. PROPOSED CROSS-LAYER DESIGN

The basic principle of our cross-layer design is to share the channel status information with the upper layers. Using this information, before sending a packet, the upper layers ideally know whether the channel gain is large enough for a successful transmission. In this situation, the upper layers proceed with the transmission as usual. On the other hand, if the channel is in a fade, the upper layers (i.e. in this paper the Medium Access Control) will suspend transmission of packets which otherwise would not be received without significant if not total loss. Transmission of such packets is bound to be lost because of a fade. It wastes power and bandwidth resources for transmission and increases packet delay. Therefore, as compared with the conventional layered approach where information is not shared across layers, our cross-layer can save bandwidth, power resources and possibly reduces packet delay.

In general, it is difficult for the transmitting node to know with certainty whether a packet will be transmitted correctly. Nevertheless, in the case of Rayleigh fading channels, one can predict the value of the channel gain based on the past values of this quantity. In other words, the receiving station can "predict" whether the next packet will be received correctly (based on the predicted value of the channel gain). This information is then relayed back to the transmitting node in the Acknowledgement packet.

Let us now concentrate on the prediction of the channel gain. It has been shown that the autocorrelation of the Rayleigh fading channel envelope can be expressed as [8]

$$\rho_\beta = \sigma^2 J_0(2\pi f_D \upsilon) \tag{5.1}$$

where f_D is the maximum Doppler spread and υ is the time shift.

Using the correlation function in (5.1), a prediction of the channel gain can be expressed as a combination of the past V measured channel gains with the range D, $[\beta(i-D)\beta(i-D-1)\ldots\beta(i-D-V+1)]$.

$$\beta_{pred}(i) = \sum_{v=0}^{V-1} a_v(i)\beta(i - v - D) \tag{5.2}$$

where $a_v(i)$, $v = 0, 1, \ldots, V - 1$ are the linear prediction coefficients and can be derived as follows

$$a(i) = \mathbf{R}^{-1}(i)\mathbf{r}(i) \tag{5.3}$$

where $\mathbf{R}(i)$ is the $V \times V$ autocorrelation matrix of the input samples, whose elements are $\mathbf{r}(i)_{v,u} = E[\beta(i - D - v)\beta^*(i - D - u)]$, $u, v = 0, 1, \ldots, V - 1$. The elements of vector $r(i)$ are

$$\mathbf{r}(i)_v = E[\beta(i)\beta^*(i - D - v)] \qquad v = 0, 1, \ldots, V - 1 \tag{5.4}$$

Readers are referred to [8] for a complete derivation of the autocorrelation function and the linear prediction coefficients.

Using the prediction, the receiving station can "determine" whether the packet will be received correctly in the next transmission. If the packet is not to be received correctly, the node performs the following actions:

- Stops the transmission of any reply packet to the sending node, freeze the Medium Access Control (MAC) until the channel gets out of the current fade.
- Notifies the sender to stop the transmission and also indicate the expected fade duration of the channel. This notification of the incoming fade can be achieved in IEEE 802.11 by setting up a special flag indicating in the header of either the Clear To Send (CTS) packet or Acknowledgment (ACK) packet. The expected average fade duration will also be stored in a field in the header. Note that we assume that the "backward channel" (from the receiver to the sender) is not itself in a fading situation during the transmission of CTS or ACK. It is a reasonable assumption since we do the prediction in advance and therefore the CTS or ACK have sufficient time to get back to the sending node before the fade starts.

Upon reception of a CTS or ACK packets containing an activated flag signalling a fade, the transmitting node performs the following actions

- Immediately halts the transmission to the destination node in the MAC, routing layer, and application layer whichever applicable.
- Obtains the expected fade duration (AFD) and schedules the transmission accordingly.

The neighbouring nodes hear this CTS, or ACK and automatically set their network allocation vectors (NAVs) accordingly. The channel can then be released to other nodes for transmission during the AFD.

Recall that F_a, the expected downtime of the channel or AFD can be expressed as [6]:

$$F_a = \frac{e^{\rho^2} - 1}{\sqrt{2\pi} f_D \rho} \tag{5.5}$$

Where:

- ρ is the ratio between the power threshold and the Root Mean Square (RMS) of the received power.
- f_D is the maximum Doppler frequency.

The basic operations involved in our cross-layer design algorithm are illustrated in Fig. 5-1.

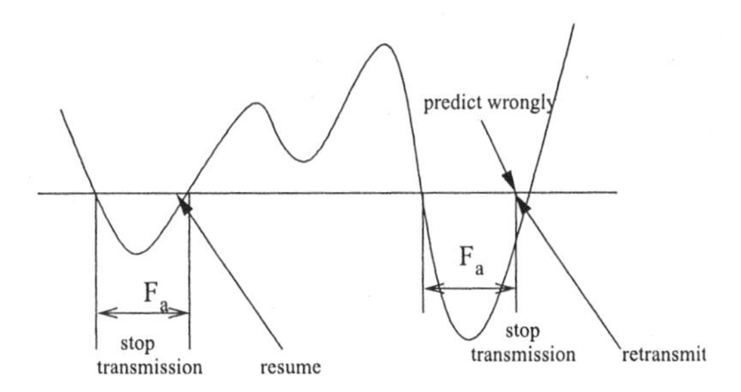

Figure 5-1. Cross-layer design using channel prediction.

2.1 Advantages and Practical Implications of the Proposed Method

It is well known that in the case of Rayleigh fading channels, the received signal can go into deep fades [6]. If the proposed cross-layer design is not used, the upper layers are not notified when the channel goes into a fade and therefore the transmitting node keeps sending packets which are discarded due to weak received power at the receiver. Therefore, the transmitting node stops receiving Acknowledgement packets from the destination node and enters the backoff state after the timeout. It then starts trying to send these packets again and if the fade is longer than the time corresponding to the specified number of allowed retries, the sending node discards the packets permanently.

Therefore, using our proposed cross-layer design approach, we can obtain the following improvements:

Firstly, it prevents the sender from unnecessary packet transmissions, which results in the reduction of power consumption for transmission. In addition, it saves bandwidth resources which can be used for other transmissions, leading to an increase in the overall network throughput. Furthermore, this cross-layer approach avoids retransmissions of packets which ultimately result in permanent packet loss. In fact, this cause of packet drop can be significant as discussed in [9].

The prediction algorithm is entirely dependent on the coefficients of the predictor which can be obtained from (5.1). However, the derivation of these values is computationally intensive and consumes a significant amount of power which may not be viable during real-time operations. We can solve this problem by pre-computing the coefficients and ''storing'' them in a file, which can be retrieved based on the Doppler frequency and the data rate. This approach is fast and energy efficient, which suits an ad hoc network environment.

One can notice that the improvements from cross-layer design are dependent on the accuracy of the prediction algorithm. In turn, the prediction mechanism is dependent on the continuing evaluation of received power values, which are obtained during packet reception. In ad hoc networks, the packet transmission is not continuous, which could result in missing received power values. This can affect the accuracy of the prediction algorithm. However, this can be alleviated by replacing actual values of the received signal by estimated values.

When the channel is in a fade, there can be no viable communication between the source and the destination. Therefore, the source and the destination will suspend the transmission for an average fade duration, F_a, which depends on the Doppler frequency and the Root Mean Square (RMS) value of the received power. However, the actual downtime of the channel during any given fade, may be greater or less than the average fade duration. If the downtime is larger than F_a, the sender resumes the packet transmission which will not be received correctly by the receiver. This may affect the overall network performance. However, as shown in the simulation results, the algorithm improves the network performance as well as reduces the packet loss and unnecessary transmissions which validates the approximation of the channel downtime by the AFD.

In order to illustrate the benefits of our cross-layer approach, we first present and discuss the simulation results.

3. SIMULATION RESULTS

In the experiment, ns2 is chosen as the simulation tool [10]. We used the standard IEEE 802.11 DCF using RTS/CTS access scheme in ns2 with the channel bandwidth of 1Mbps. We analyse our prediction algorithm and cross-layer design for several scenarios.

3.1 Channel Prediction Accuracy

Firstly, we analyse how accurate is the prediction algorithm in the simulation environment. There are two mobile nodes A and B being separated by a distance 120 meters. Node B is moving away from node A with velocity 1 m/s. Node A transmits a constant number of packets to node B with the rate of 50 packets/second, each of size 512 bytes. The predicted received power of the next packet is then compared against actual received power. This result is shown in Fig. 5-2. As shown, the algorithm does predict accurately future values of the received power. However, it is noticed that this is a relatively high data rate in ad hoc networks. It is also known that the accuracy of the prediction algorithm is proportional to the data rate. Therefore, it is interesting to find out at what rate the prediction is still reasonably accurate. By varying the packet rate, we have been able to conclude that with a mobility of 1 m/s, the prediction algorithm is no longer accurate at approximately 5 packets/second.

It is also important to emphasise that, in this study we are interested in the relative mobile speeds not the absolute speeds; two mobile nodes can have a relative speed of 1 m/s while themselves travelling at very high speeds. It is interesting to notice that our cross layer design is applicable mostly for low relative speeds which also corresponds to cases where improving the link reliability is more interesting; a low relative speed between two nodes means that those two nodes are moving in similar directions and hence, it is likely that the link between them will last for a long time. A timeout due to multipath fading, leading to an establishment of a new route, could degrade the performance of such a link. Hence it can be concluded that this cross layer design is applicable to low relative speeds, where it is likely to have long lasting links connectivity between moving nodes.

3.2 Four-Node Scenario

In this scenario, we study how the cross layer mechanism performs with the presence of channel contention as well as fades. As mobile nodes compete for one channel, the probability of packet loss due to contention is higher. Our proposed cross-layer design alleviates this by minimizing the number of retransmissions. Therefore, it is predicted that the number of packet loss due to channel contention will be lower using our scheme. In addition, we allow for a receiver to release the channel for other competing nodes when it predicts a fade. Therefore, we can expect an improvement in the average connection throughput.

The simulation scenario is shown in Fig. 5-3. In this scenario, mobile nodes 1, 2, 3, and 4 move away from mobile node 0 with a speed of 1 m/s in the designated directions. Node 0 transmits to node 1, 2, 3 and 4 with

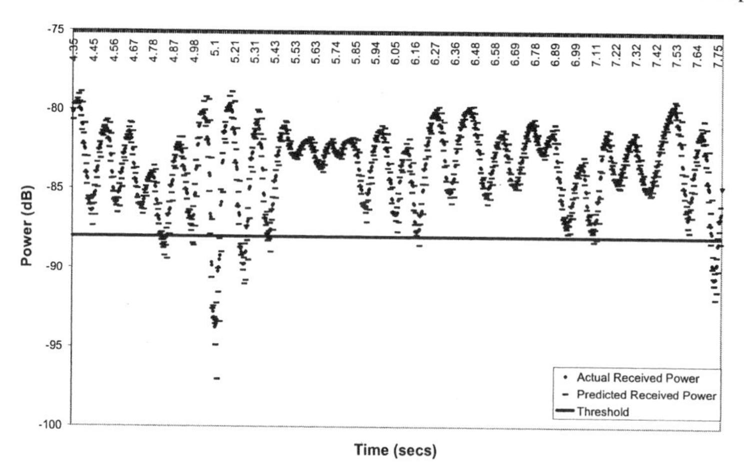

Figure 5-2. The Comparison Between Predicted and Actual Received Power

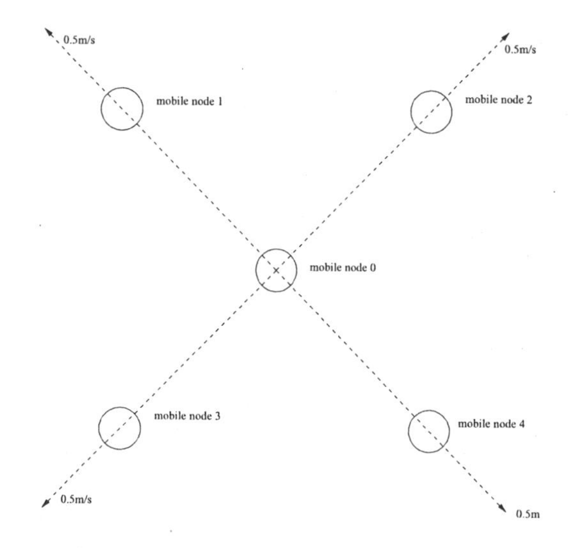

Figure 5-3. Diagram of Four-Nodes Scenario

a constant packet rate with the same size of 512 bytes. As predicted, we observe significant reductions in control packets, packet drops due to channel contention and packet drops due to low power as shown in Fig. 5-4, Fig. 5-5 and Fig. 5-6 respectively. Specifically, it can be seen that the cross-layer design achieves between 20-30% reduction in control packet as in Fig. 5-4 and a huge reduction in packet drops due to low power as in Fig. 5-6. This is the main merit of using the new cross-layer design. Through reducing unnecessary

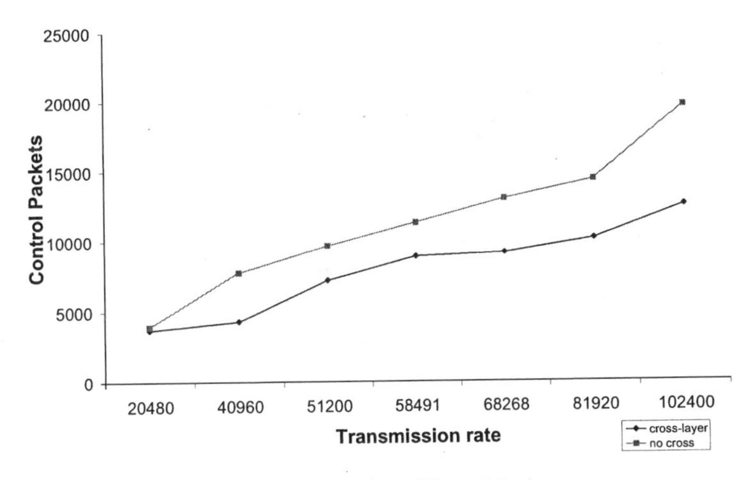

Figure 5-4. Number of Control Packets

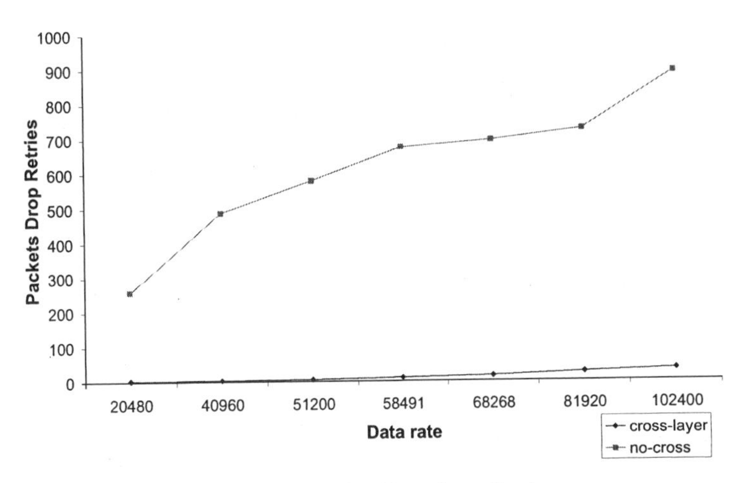

Figure 5-5. Packets Drop due to Retries

packet transmissions, it saves bandwidth resources, power consumptions and reduces packet loss due to contention. As a result, the cross-layer approach achieves a better network throughput as noted in Fig. 5-7.

3.3 General Random Mobility Model

We intend to measure the average improvement of using our cross-layer design in a scenario which is conventionally adopted by many researchers. In this scenario, there are 50 mobile nodes moving inside a square of 1000x1000

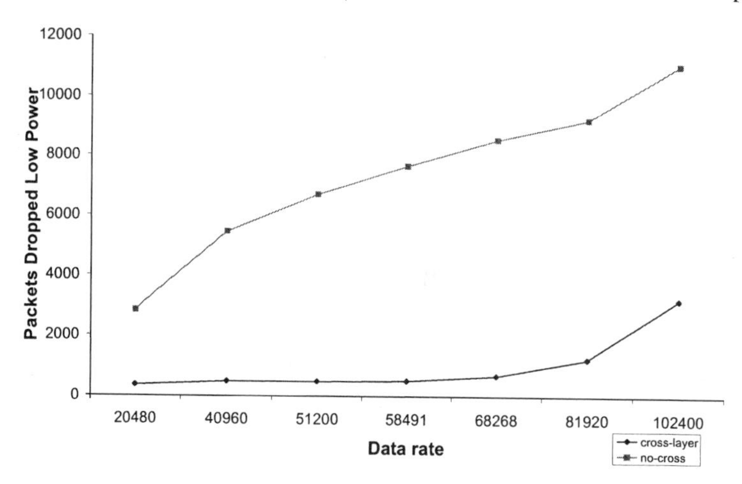

Figure 5-6. Packets Drop due to Low Power

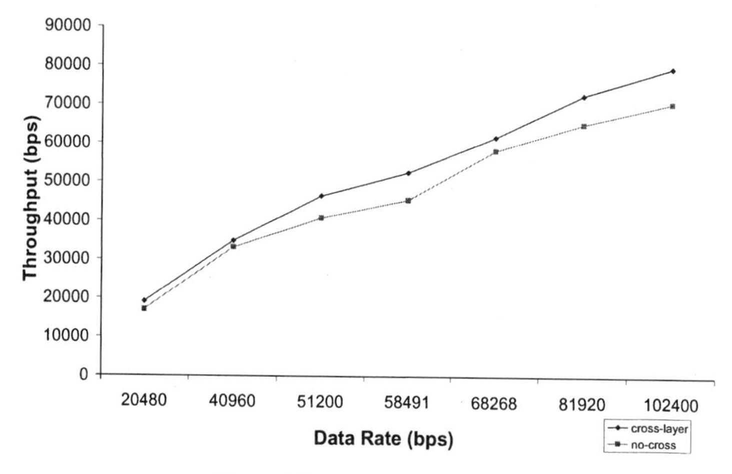

Figure 5-7. Average Throughput

meters. Each mobile node moves to a random destination according to random waypoint mobility model [11] with a speed of 1 m/s. After reaching the destination, the mobile node pauses for 10 seconds before moving to another random position. On average, there are 20 simultaneous active data connections. Data packets are of a same size 512 bytes. The average network throughput is presented in Fig. 5-8. One may notice that the improvement in network throughput is insignificant at low data rate. This is because the prediction algorithm does not accurately predict the received power. However, at normal and high data rate, due to a better prediction, the cross-layer approach

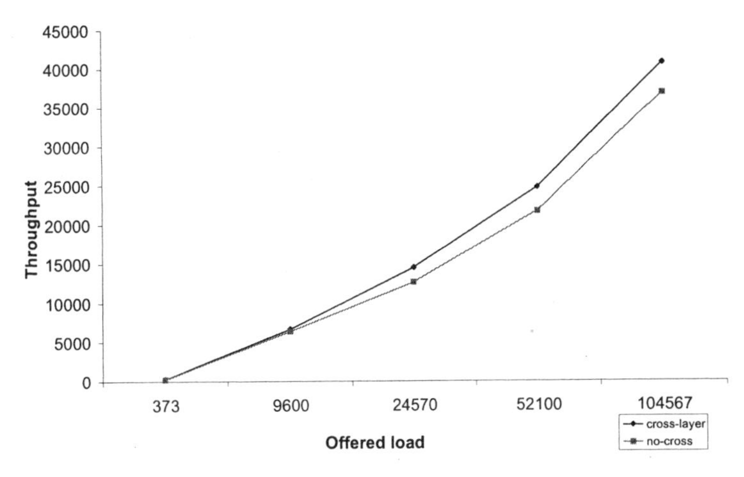

Figure 5-8. Throughput for 50 node simulation

increases the network throughput by 10%. The improvement is noticeable from 24570 bps or 6 packets/seconds. Once again, it is noticed that this result is consistent with the results from Section 3.1. This is a significant result considering that this is an "all-win" design because we do not sacrifice any resource for the performance improvement.

4. MARKOVIAN MODEL FOR RAYLEIGH FADING CHANNEL

Several analytical models have been proposed for the Rayleigh channel [12--16]. In [15], a Markov process was derived for the channel gain, which was defined as the ratio between the received signal power and transmitted signal power. In [14], a Markov model for the signal to noise ratio was proposed and then theoretically analysed. This work was later expanded for slow Rayleigh fading channels [12]. In [16], it introduced the concept of using a Markov model to represent the "good" and "bad" states of a Rayleigh channel. However, it did not present any theoretical derivation of the probability of the channel being "good" or "bad".

In this section, we prove that we can use a Markov model consisting of two states. Each state is respectively associated with the probability for the received signal to be below or above a given threshold. The following parameters are used in the analysis.

Table 5-1. Summary of Parameters for Rayleigh Channels

Parameters	*Definitions*
P	Received signal
Nr	Level Crossing Rate
$p(r)$	Prob of density function for P
F_a	Average fade duration
P_{th}	Received power threshold
f_D	Maximum Doppler frequency
ρ	The ratio P_{th}/P_{rms}
P_{rms}	The RMS value for the received power
$t_{i,j}$	Transition probability from state i to state j
S_0	"bad" state corresponding to a fade
S_1	"good" state (channel not in a fade)
π_0	The stationary probability at state S_0
π_1	The stationary probability at state S_1
ι	The RMS value of received signal before envelope detection

The probability density function (p.d.f.) of P can be expressed as

$$p(r) = \begin{cases} \frac{r}{\iota^2} exp\left(-\frac{r^2}{2\iota^2}\right) & (0 \leq r \leq \infty) \\ 0 & (r < 0) \end{cases} \tag{5.6}$$

where ι is the RMS value of the received signal before envelope detection [6].

The RMS value of the received power can be derived as

$$P_{rms} = \sqrt{E[r^2]} = \sqrt{\int_0^\infty r^2 p(r) dr} = \sqrt{2}\iota \tag{5.7}$$

Therefore

$$\rho = \frac{P_{th}}{P_{rms}} = \frac{P_{th}}{\sqrt{2}\iota} \tag{5.8}$$

where f_D is the maximum Doppler frequency. The number of times per second the received power P crosses the level P_{th} in a positive direction is

$$N_r = 2\pi f_D \rho e^{\rho^2} \tag{5.9}$$

As briefly mentioned previously, we divide the received power into two levels:

$$\text{level 1:} \quad 0 < P \leq P_{th} \tag{5.10}$$

$$\text{level 2:} \quad P_{th} < P < \infty \tag{5.11}$$

The Rayleigh channel is either in state S_1 if $P_{th} < P < \infty$ or S_0 if $0 < P \leq P_{th}$

With the p.d.f. of the received signal as in (5.6) the steady state probabilities p_1 and p_2 of the state S_1 and S_0 are

$$\pi_1 = Pr(r > P_{th}) = \int_{P_{th}}^{\infty} \frac{r}{\iota^2} exp\left(-\frac{r^2}{2\iota^2}\right) dr = e^{-\rho^2} \tag{5.12}$$

$$\pi_0 = Pr(0 < r \leq P_{th}) = \int_{0}^{P_{th}} \frac{r}{\iota^2} exp\left(-\frac{r^2}{2\iota^2}\right) dr = 1 - e^{-\rho^2} \tag{5.13}$$

In order to calculate the transition probabilities $t_{i,j} \quad 0 \leq i, j \leq 1$, it is noticed that in 1 second, there are N_r crossings from state S_0 to state S_1. Therefore, the duration between each crossing is $1/N_r$. Furthermore, the channel remains in fade (i.e. in state S_0) for F_a seconds. Therefore, the transition probability $t_{0,1}$ and $t_{1,0}$ can be approximated as:

$$t_{1,0} = F_a/(1/N_r) = e^{-\rho^2}(e^{\rho^2} - 1) = 1 - e^{-\rho^2} \tag{5.14}$$

$$t_{0,1} = \frac{1/N_r - F_a}{1/N_r} = 1 - N_r F_a = 1 - e^{-\rho^2}(e^{\rho^2} - 1) = e^{-\rho^2} \tag{5.15}$$

Therefore, we can conclude that the state equilibrium equation holds.

In summary, we can use a Markov model to model the "good" and "bad" states of a Rayleigh channel. The channel is in "good" state S_1 with probability $e^{-\rho^2}$ with expected duration $1/(\sqrt{2\pi} f_D \rho)$ and in "bad" state S_0 with probability $1 - e^{\rho^2}$ with expected duration $(e^{\rho^2} - 1)/(\rho f_D \sqrt{2\pi})$. Fig. 5-9 illustrates the dynamics of the Rayleigh fading channel.

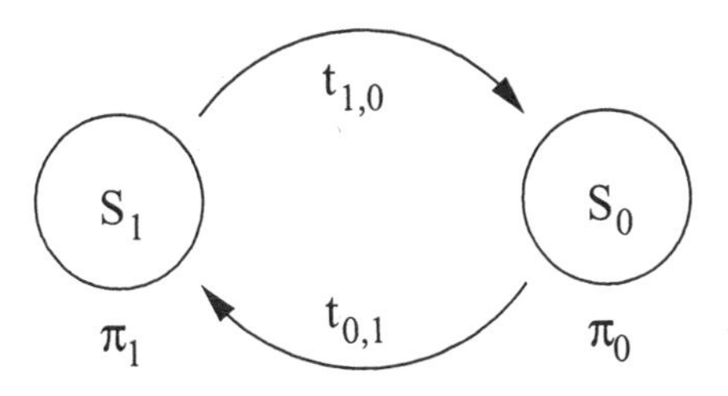

Figure 5-9. The Channel Model

5. IEEE 802.11 MARKOVIAN MODEL

In this section, we study the performance of IEEE 802.11 DCF under a Rayleigh fading channel with and without the cross-layer approach. In this paper, we use the results from [17]. The work in [17] was related to proposing a new comprehensive Markov model for backoff stages of a station using IEEE 802.11 DCF. However, this work is based on a perfect channel assumption.

Therefore, we need to selectively adapt these results, taking into account the characteristics of the Rayleigh fading channel. For convenience, the Markov model in [17] is reproduced in Fig. 5-10 with the parameters being summarized in Table 5-2.

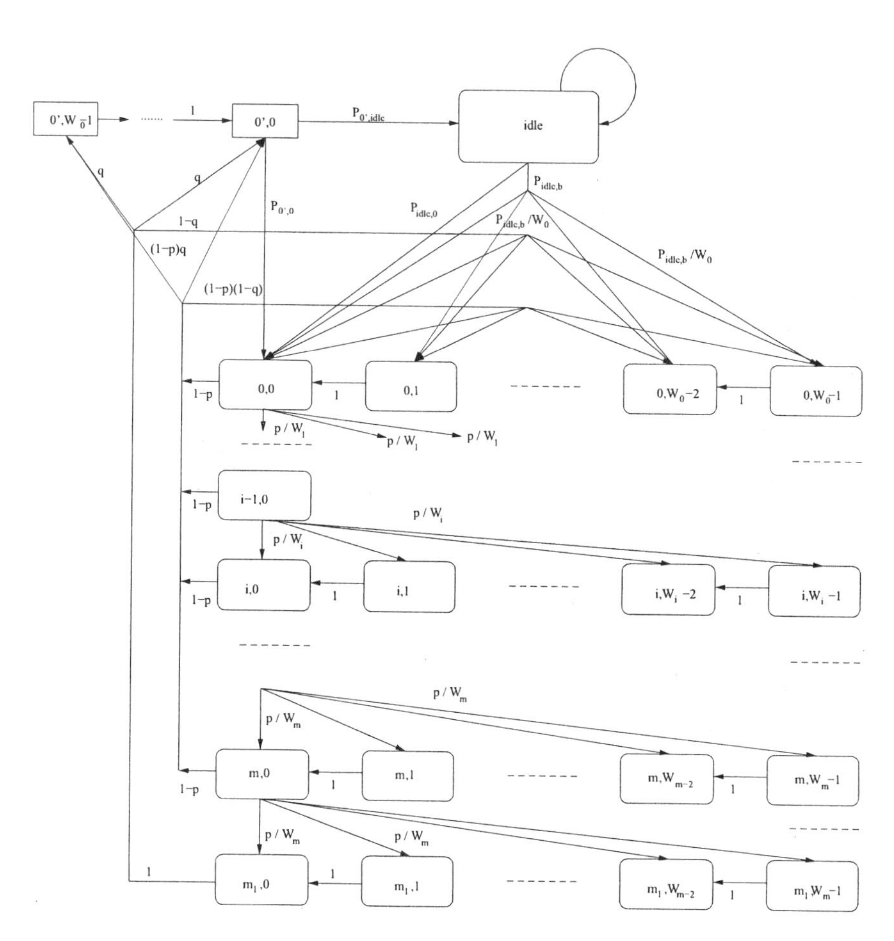

Figure 5-10. Markov Model for Backoff of IEEE 802.11

Hereon, for a better distinction between with/without the cross-layer approach. The parameters are appended with the subscript $_c$ (cross-layer) for cross-layer approach and $_{nc}$ (no-cross layer) for normal approach unless they have the same values for both cases.

Table 5-2. Summary of Parameters for IEEE 802.11 Markovian Model

Parameters	*Definitions*
p	Prob of a packet not received correctly
$b(i, j)$	Prob of the node in backoff stage $S_{i,j}$
$b(idle, 0)$	Stationary prob at idle state $S_{idle,0}$
W_i	Backoff window size at stage $s(t) = i$
W	Min backoff window size
q	Prob of the queue is empty
$P_{tr(n-1)}$	Prob of at least one out of $n - 1$ node transmits
$P_{s(n-1)}$	Prob of one out of $n - 1$ transmit successfully
τ	Prob of the packet to transmit
$S_{i,j}$	Backoff state of mobile node when $s(t) = i$ and $b(t) = j$
λ	Average packet arrival rate
$P_{0',0}$	Transition prob from state $S_{0',0}$ to $S_{0,0}$
$P_{idle,0}$	Transition prob from state $S_{idle,0}$ to $S_{0,0}$
$P_{idle,b}$	Transition prob from state $S_{idle,0}$ to back-off state
$P_{0',idle}$	Transition prob from state $S_{0',0}$ to $S_{idle,0}$
$\bar{\sigma}$	Average channel slot time
$P_{tr(n)}$	Prob at least one node out of n nodes transmit
$P_{s(n)}$	Prob successful transmission for n nodes
$P_{tr(n-1)}$	Prob at least out of $n - 1$ nodes transmit
$P_{s(n-1)}$	Prob successful transmission for $n - 1$ nodes
U	Channel throughput
Q_l	Queue length
D_{ave}	Average delay of a packet
P_l	Probability of packet lost
f_D	Doppler frequency
μ	Packet processing rate
σ	Channel idle slot (system slot)
δ	Propagation delay
$S_{idle,0}$	State channel is idle
$\bar{\sigma}_s$	Average slot time at saturation
m_1	The last backoff stage where packet may be discarded
m	The number of backoff stages prior to discarding the data packet
η	Queue utilization for deriving the average delay

5.1 Model Without Cross-layer Design

In comparison to the normal channel, the packet transmission under Rayleigh channel exhibits some differences. Firstly, a packet is discarded either when there is a collision or when the channel is "bad". Therefore, the probability of discarding packets under the Rayleigh channel is higher. Furthermore, a successful RTS/CTS exchange does not necessarily guarantee a successful data packet transmission because the channel may still turn "bad" after receiving the CTS. Therefore, T_c under normal channel changes to:

$$T_{c_{nc}} = \frac{p(\text{RTS+DIFS} + \delta) + (1-p)\pi_0(\text{RTS+2SIFS} + 2\delta + \text{CTS} + \text{H+P}_\text{L})}{p + (1-p)\pi_0} \tag{5.16}$$

where π_0 is the probability of the Rayleigh channel in "bad" state and is derived using (5.13).

5.1.1 Derivation of τ_{nc} and p_{nc}.

The packet can be destroyed either due to collision or due to low received power, i.e. when the received power is lower than threshold.

$$p_{nc} = 1 - (1-\tau_{nc})^{n-1} + \pi_0(1-\tau_{nc})^{n-1} \tag{5.17}$$

From (5.17), we can rewrite of the probability of packet transmission as

$$\tau_{nc} = 1 - \left(\frac{1-p_{nc}}{1-\pi_0}\right)^{1/(n-1)} \tag{5.18}$$

where π_0 is denoted as the probability of the channel in the "bad" state and is derived in (5.13).

We also know from [17] that the probability for transmitting a packet can be expressed as

$$\tau_{nc} = \sum_{i=0}^{m+1} b(i,0)_{nc} = \frac{b(0,0)_{nc}(1-p_{nc}^{m+2})}{1-p_{nc}} \tag{5.19}$$

where m is defined as the number of backoff stages for a station before discarding the packet. Typically, for IEEE 802.11 b using Direct Sequence Spread Spectrum (DSSS), $m = 5$.

Using (5.19) and (5.18), we can obtain τ_{nc} and hence p_{nc}.

5.1.2 Channel Throughput Analysis.

The probability for a successful transmission in a given channel slot is equal to the probability of exactly one mobile node transmitting on the channel, i.e. $\tau_{nc}(1-\tau_{nc})^{n-1}$. Since there are n combinations for n nodes, the probability of a packet transmission in a channel slot is:

$$P_{success_{nc}} = n\tau_{nc}(1-\tau_{nc})^{n-1} \tag{5.20}$$

Let U_{nc} be the normalized channel throughput which is defined as the fraction of channel time being used for transmitting data bits. The actual format of a data packet is shown in Fig. 5-11. It consists of actual payload P_L and header information H. Therefore, the channel throughput can be expressed as

$$U_{nc} = \frac{n\tau_{nc}(1-\tau_{nc})^{n-1}P_L}{\bar{\sigma}_{nc}} \tag{5.21}$$

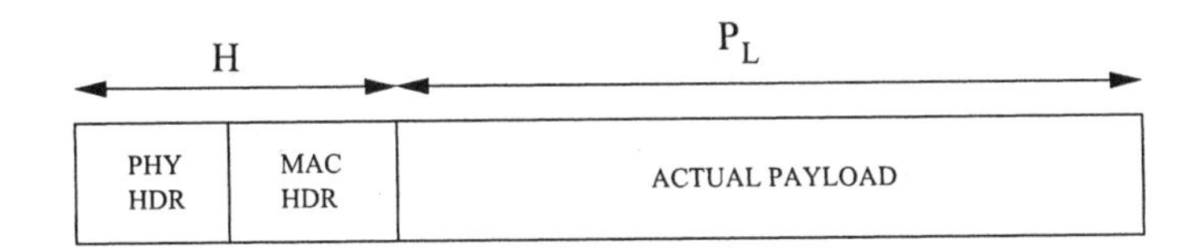

Figure 5-11. Data Packet Format

5.1.3 Packet Processing Rate.

In order to derive the packet processing rate, it is assumed that incoming packets are of the *same size*, arriving exponentially with the rate λ. Assuming μ_{nc} as the packet processing rate, and $\bar{\sigma}_{nc}$ as the average slot time, the probability of successfully transmitting a packet in a slot time is $\tau_{nc}(1-\tau_{nc})^{n-1}$. Therefore the processing rate is:

$$\mu_{nc} = \tau_{nc}(1-\tau_{nc})^{n-1}/\bar{\sigma}_{nc} \tag{5.22}$$

5.1.4 Packet Loss Analysis.

A data packet can be dropped when the sending queue is full (queue loss) or after a certain number of retries (contention loss). It can be seen from the Markov model in Fig. 5-10, the state $S_{m_1,0}$ is the state where the node drops the RTS packet and the DATA packet, resets the backoff window and enters a new backoff state. Therefore, the probability of packet loss due to contention is:

$$P_{lc_{nc}} = b_{nc}(m_1, 0) = p_{nc}^{m+1} b_{nc}(0,0) \tag{5.23}$$

It is reminded that $b_{nc}(m_1, 0)$ is the final moment at the final backoff stage where the packet is discarded after a predefined number of retries (Short Retry Limit as specified in the standard).

Using an arrival rate of λ for packets which are dropped with probability $P_{lc_{nc}}$, the processing rate being μ_{nc} and the queue length Q_l, we deduct that the probability of a packet being destroyed by the queue overflow is

$$P_{lq_{nc}} = \left(\frac{\lambda(1-P_{lc_{nc}})}{\mu_{nc}}\right)^{Q_l} \left(\frac{1-(\lambda(1-P_{lc_{nc}})/\mu_{nc})}{1-(\lambda(1-P_{lc_{nc}})/\mu_{nc})^{Q_l+1}}\right) \tag{5.24}$$

Therefore the total probability of packet loss is:

$$P_{l_{nc}} = \left(\frac{\lambda(1-P_{lc_{nc}})}{\mu_{nc}}\right)^{Q_l} \left(\frac{1-(\lambda(1-P_{lc_{nc}})/\mu_{nc})}{1-(\lambda(1-P_{lc_{nc}})/\mu_{nc})^{Q_l+1}}\right) + p_{nc}^{m+1} b_{nc}(0,0) \tag{5.25}$$

5.1.5 Delay Analysis.

We use queuing theory to derive the average delay for a successfully transmitted packet. The effective arrival rate of the data packet is

$$\lambda_{eff_{nc}} = \lambda\left[1 - p_c^{m+1}b_c(0,0)\pi_1 - \pi_1\left(\frac{\lambda(1-P_{lc_c})}{\mu_c}\right)^{Q_l} X\right] \quad (5.26)$$

where $X = \frac{1-(\lambda(1-P_{lc_c})/\mu_c)}{1-(\lambda(1-P_{lc_c})/\mu_c)^{Q_l+1}}$

From [18], using $M/M/1/Q_l$ queue results, the average number of packets in the system is

$$L_{nc} = \frac{\eta_{nc}(1-(Q_l+1)\eta_{nc}^{Q_l} + Q_l\eta_{nc}^{Q_l+1}\eta_{nc})}{(1-\eta_{nc})(1-\eta_{nc}^{Q_l+1})} \quad \text{where} \quad \eta_{nc} = \frac{\lambda_{eff_{nc}}}{\mu_{nc}} \quad (5.27)$$

Therefore using Little theorem [18], the average delay for a packet is

$$D_{ave_{nc}} = \frac{L_{nc}}{\lambda_{eff_{nc}}} \quad (5.28)$$

5.2 Model With Cross-layer Design

Using the cross-layer approach, assuming a perfect prediction, the source suspends its transmission until the channel is back to good state. Therefore, effectively, there is no packet loss due to low power and the value ofT_c is calculated as:

$$T_{c_c} = \text{RTS} + \text{DIFS} + \delta \quad (5.29)$$

5.2.1 Derivation of τ_c and p_c.

Assume that a packet drop is only due to collision. The packet is destroyed when two nodes out of n nodes transmit at the same time. In other words, the probability of collision is equal to the probability of at least one out of$n-1$ station transmitting.

$$p_c = 1 - (1-\tau_c)^{n-1} \quad (5.30)$$

From (5.30), we can deduct that the probability of a mobile node transmitting is

$$\tau_c = 1 - (1-p_c)^{1/(n-1)} \quad (5.31)$$

From [17], the probability for transmitting a packet can be expressed as

$$\tau_c = \sum_{i=0}^{m_1} b(i,0)_c = \frac{b(0,0)_c(1-p_c^{m+2})}{1-p_c} \quad (5.32)$$

Using (5.32) and (5.31), τ_c and hence p_c are readily obtained.

5.2.2 Channel Throughput Analysis.

Using our cross-layer design, when the channel is "bad", the source stops sending packets and only resumes the transmission when the channel gets out of the fade. Therefore, there is effectively no packet drop due to low power. It is also recalled that, on average the channel is in "bad" state S_0 with probability π_0 and in "good" state (state S_1) with probability π_1. Using the same derivation as in the previous section, the actual channel throughput is

$$U_c = \frac{n\tau_c(1-\tau_c)^{n-1}P_L}{\bar{\sigma}_c} \times \pi_1 \tag{5.33}$$

where π_1 can be obtained from (5.12).

5.2.3 Packet Processing Rate.

Using the same argument as in the previous section, the packet processing rate is

$$\mu_c = \tau_c(1-\tau_c)^{n-1}/\bar{\sigma}_c \tag{5.34}$$

5.2.4 Packet Loss Analysis.

We use a similar derivation as in the previous section, noting the difference in the channel uptime. The packet loss due to contention can be expressed as:

$$P_{lc_c} = p_c^{m+1} b_c(0,0)\pi_1 \tag{5.35}$$

The packet loss due to queue overflow is

$$P_{lq_c} = \pi_1 \left(\frac{\lambda(1-P_{lc_c})}{\mu_c}\right)^{Q_l} \left(\frac{1-(\lambda(1-P_{lc_c})/\mu_c)}{1-(\lambda(1-P_{lc_c})/\mu_c)^{Q_l+1}}\right) \tag{5.36}$$

Therefore, the total probability of packet loss is

$$P_{l_c} = p_c^{m+1} b_c(0,0)\pi_1 + \pi_1 \left(\frac{\lambda(1-P_{lc_c})}{\mu_c}\right)^{Q_l} \left(\frac{1-(\lambda(1-P_{lc_c})/\mu_c)}{1-(\lambda(1-P_{lc_c})/\mu_c)^{Q_l+1}}\right) \tag{5.37}$$

5.2.5 Delay Analysis.

Since the source suspends the packet transmission during the channel downtime, the total delay must be scaled by $1/\pi_1$, i.e. :

$$D_{ave_c} = \frac{L_c}{\pi_1 \lambda_{eff_c}} \tag{5.38}$$

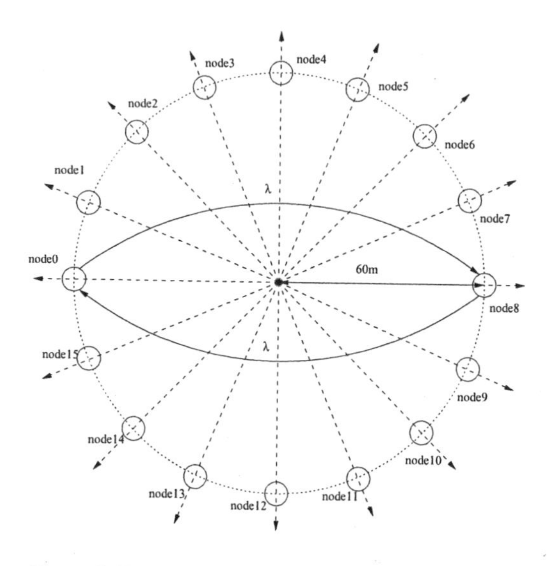

Figure 5-12. Scenario for Analytical Model Validation

Where

- $\lambda_{eff_c} = \lambda - P_{lq_c}\lambda$
  $= \lambda\left[1 - p_c^{m+1} b_c(0,0)\pi_1 - \pi_1 \left(\frac{\lambda(1-P_{lc_c})}{\mu_c}\right)^{Q_l} \left(\frac{1-(\lambda(1-P_{lc_c})/\mu_c)}{1-(\lambda(1-P_{lc_c})/\mu_c)^{Q_l+1}}\right)\right]$
- $\eta_c = \frac{\lambda_{eff_c}}{\mu_c}$
- $L_c = \frac{\eta_c(1-(Q_l+1)\eta_c^{Q_l}+Q_l\eta_c^{Q_l+1}\eta_c)}{(1-\eta_c)(1-\eta_c^{Q_l+1})}$
- π_1 can be derived from (5.12)

6. VALIDATING ANALYTICAL MODELS

In this section, we present the results from the analytical model. We then compare the theoretical results against the simulation values. The scenario for both theoretical and simulation analysis is shown in Fig. 5-12. In this scenario, 16 mobile nodes are equally located on a circle with a radius of 50 m. Each node moves away from the centre with the velocity 0.5 m/s on the designated direction. Therefore, the relative speed between each source-destination pair is 1m/s. Each transmits its opposite with a constant packet rate and with a constant packet size of 1000 bytes. The transmitting queue length for each mobile node is set to 50 packets. The simulation time is 2 minutes.

The values of theoretical connection throughput and simulation connection throughput when using and not using the cross-layer approach are shown in Fig. 5-13. Firstly, we notice a common trend of the theoretical and the simulation values at low data rate. In fact there is a close match between the theoretical results and the analytical results. However, at high data rate, the theoretical model achieves a better result. It is important to emphasize that the difference between the theoretical results and simulation results was also independently observed in [19, 20]. At high data rate, under increasing presence of packet collisions due the contention, IEEE 802.11 DCF exhibits some instability [19] due to stations' backoffs. Furthermore, the current backoff mechanism in IEEE 802.11 does not effectively schedule the retransmissions under such condition. As a result, IEEE 801.11 does not achieve the theoretical bounds.

It is also interesting to note that at low data rate IEEE 802.11 does manage to transport the packet load very well. It is reflected by the close coherence between the channel throughput and its subjected packet load. For example, at packet arrival rate of 4 packets/sec, the channel throughput is approximately 0.50. At this packet rate, the total packet load is 4*16*8000 = 512 kpbs. If IEEE 802.11 manages to transfer all the load, the channel throughput should be 0.512. As shown, the simulation results show a good match of 0.507 and the theoretical results show a close match of 0.484.

Moreover, at low data rate, using cross-layer design will result in a lower channel throughput as the prediction algorithm does not receive enough samples to achieve an accurate prediction. From Fig. 5-13, it can be seen that the cross-layer design starts to gain performance improvement when the packet rate is approximately between 5-6 packets/second. This is once again coherent with the findings in Section 3.1. At higher data rate, it is noticed that the cross-layer design does achieve a better connection throughput which is confirmed by the simulation results. However, at very high data rate, the performance of our cross layer scheme is limited by the problem of queue overflow.

The theoretical and simulation average packet delays are shown in Fig. 5-14. It is noted that there is a good match between the theoretical values and the simulation values. At the traffic rate higher than 8 packets/sec, the actual load is higher than the maximum packet processing rate. As the model for deriving the average packet delay using queuing model is not valid in this region, we have only presented analytical delay values up to 8 packs/sec.

7. CONCLUSION

In summary, we have presented a new cross-layer approach to improve the performance of ad hoc networks under a Rayleigh fading channel. The approach, using the predictability of the channel, feeds the channel status back to the upper network layers. The upper layers can then suspend the packet transmission when the channel enters a fade. This minimizes

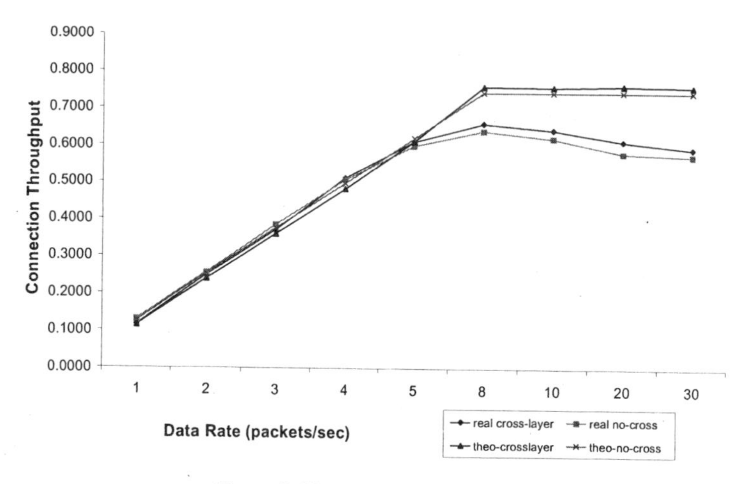

Figure 5-13. Network Throughput

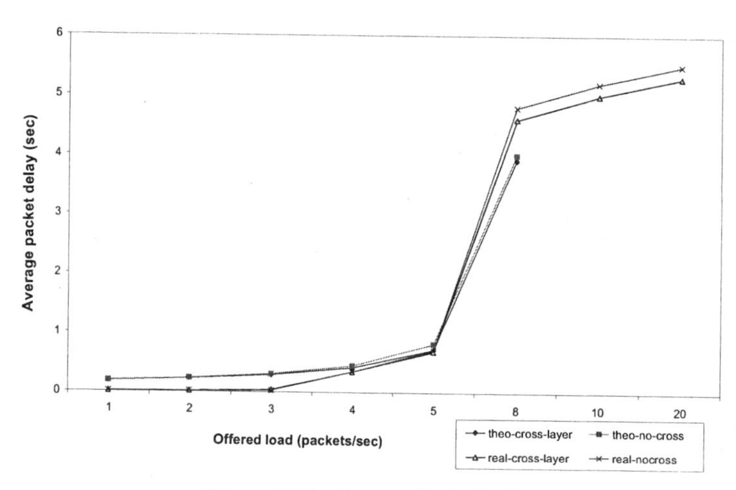

Figure 5-14. Average Packet Delay

unnecessary packet retransmissions, saves power and bandwidth usage and therefore increases the connection throughput by approximately 5-10%.

In this chapter, we have also proposed a two Markovian models: one for modelling the uptime and downtime of the Rayleigh fading channel and the other one for analyzing the performance of IEEE 802.11 DCF using the RTS/CTS handshake. We have then used these two Markovian models to theoretically compare the performance of the IEEE 802.11 DCF when using or not using our cross layer approach. This analytical model has been validated against simulations confirming the accuracy of our theoretical models. It is expected that this model will be particularly interesting to predict when or if our cross layer design approach will be useful, depending on a large set of parameters such as relative speed, packet rate transmission frequency, node density etc. Also, an interesting direction for this work on cross layer design would be to study how it could impact on the TCP performance of multihop wireless networks.

REFERENCES

1. M. Takai and J. Martin, ''Effects of wireless physical layer modeling in mobile ad hoc networks.'' [Online]. Available: citeseer.nj.nec.com/460094.html
2. A. J. Goldsmith and S. B. Wicker, ''Design challenges for energy-constrained ad hoc wireless networks,'' *IEEE Wireless Communications*, vol. 02, 2002.
3. Y. Fang and A. B. McDonald, ''Cross-layer performance effects of path-coupling in wireless ad hoc networks: Implications for throughput, power and scalability,'' in *IEEE IPCCC*, 2002.
4. S. Toumpis and A. Goldsmith, ''Performance, optimization, and cross-layer design of media access protocols for wireless ad hoc networks,'' in *IEEE ICC*, 2003.
5. W. Yuen, H. Lee, and T. Andersen, ''A simple and effective cross layer networking system for mobile ad hoc networks,'' 2002. [Online]. Available: citeseer.nj.nec.com/yuen02simple.html
6. T. S. Rappaport, *Wireless Communication: Principles and Practice.* Printice Hall, 1996.
7. A. Muqattash and M. Krunz, ''Cdma-based mac protocol for wireless ad hoc networks,'' in *MobiHoc 2003*, Annapolis, MaryLand, USA, June 2003, pp. 153--164.
8. A. Kurniawan, ''Predictive power control in cdma systems,'' Ph.D. dissertation, University of South Australia, 2003.
9. Z. F. et al, ''On tcp performance in multihop wireless networks,'' in *UCLA WiNG Technical Report*, 2002.
10. ''Ns2,'' Internet Website. [Online]. Available: http://www.isi.edu/nsnam
11. D. Johnson and D. Maltz, ''Dynamic souce routing in ad hoc wireless networks,'' in *Mobile Computing*, 1996.
12. C. C. Tan and N. C. Beaulieu, ''On first-order markov modeling for the rayleigh fading channel,'' *IEEE Transactions on Communications*, 2000.

13. H. S. Wang and P. Chang, ''On verifying the first-order markovian assumption for a rayleigh fading channels,'' *IEEE Transaction on Vehicular Technology*, 1996.

14. H. S. Wang and N. Moayeri, ''Modeling, capacity, and joint source/channel coding for rayleigh fading channels,'' in *IEEE Vehicular Technology Conference (VTC)*, 1993.

15. R. Chen, K. C. Chua, B. T. Tan, and C. S. Ng, ''Adaptive error coding using channel prediction,'' *Wireless Networks*, vol. 5, no. 1, pp. 23--32, 1999.

16. S. R. A. A. Abouzeid and M. Azizoglu, ''Stochastic modeling of tcp over lossy links,'' in *INFOCOM*, 2000.

17. P. P. Pham, ''Comprihensive analysis of ieee 802.11,'' Institute for Telecommunications Research Internal Report, October 2003. [Online]. Available: http://www.itr.unisa.edu.au/ppham/ITRreport/report.pdf

18. L. Kleinrock, *Queueing systems.* New York : Wiley, 1975-1976, ch. 3, p. 120.

19. G. Bianchi, ''Performance analysis of the ieee 802.11 dcf,'' *JSAC*, vol. 18, no. 3, pp. 535--547, March 2000.

20. H. S. Chhaya and S. Gupta, ''Performance modeling of asynchronous data transfer methods of ieee 802.11 mac protocol,'' *Wireless Networks*, vol. 3, no. 3, pp. 217--234, 1997.

PART 2:

IDEAS FOR ADVANCED MOBILITY SUPPORT

Chapter 6

FEDERATED SERVICE PLATFORM SOLUTIONS FOR HETEROGENEOUS WIRELESS NETWORKS

Herma van Kranenburg[1], Ronald van Eijk[1], Mortaza S. Bargh[1] and Jacco Brok[2]

[1]Telematica Instituut, P.O. Box 589, Enschede, 7500 AN, The Netherlands; [2]Lucent Technologies, Bell Labs, Capitool 5, Enschede, 7521 PL, The Netherlands

Abstract: Provisioning of mobile services that are tailored to the needs, capabilities and (network) environment of users, and are available and maintained in a seamless manner when users are roaming (or changing terminal), is an ambition that requires advanced support functions. We studied such support functions for seamless mobility and tailored IP-connectivity by a federated service platform approach: interoperability issues between the various (network) technologies and administrative domains are solved through cooperation of distributed software components in the service control layer that can run on different hardware (in networks and on terminals) operating in different administrative domains. We have built Service Platform components for the server and client-side to facilitate mobile users equipped with terminals with multi-mode interfaces. Federation between various Service Platform operators and forwarding user credentials enables user or terminal to authenticate on various access networks. This ensures that consistent and coherent service access is provided, independent of the current network domain. The current chapter describes (conceptual and validated) solutions for access network selection, seamless handovers, and authentication in a WLAN/GPRS/Ethernet environment with focus on the federated Service Platform, the client-side, and privacy and trust relationships.

Key words: seamless roaming, access network detection and selection, heterogeneous environment, 4G, authentication, handover, service platform, federation, mobility management, session control

1. INTRODUCTION

Based on a previous paper [1] and our project 4GPLUS [2], this chapter describes our approach, vision, solutions and results on facilitating software infrastructures enabling seamless roaming in an integrated environment of heterogeneous networks, services and devices.

Common reality is a world with multiple platforms based on different technologies (such as Wireless LAN, fixed, 2.5 G and 3G mobile networks) present in different domains (at home, at work, in public areas) and operated by different administrators. Users typically have a few subscriptions with (network) operators and Application Service Providers (ASPs) and they dispose of a variety of terminals. We call the heterogeneous integrated environment of various networks, services and terminals "4G environment". We envision that in a 4G environment seamless integration of fixed, wireless and mobile networks will be the standard and that user centred applications, tailored to the needs, capabilities and (network) environment of users will be the norm. Moreover we expect that session portability across network interfaces is required where the session is maintained in a - for the user and application provider - transparent yet configurable way regardless of the end-user's point of attachment, while at the same time the service will adapt to the new environmental capabilities and resources. This puts flexible technological requirements on the facilitating service framework. We research key aspects of service control, see Fig. 6-1, and develop a federated Service Platform (SP) framework that enables seamless roaming of mobile users between heterogeneous networks. Our conceptual and validated SP solutions meet requirements of seamless mobility, generic service/network access, and session control, as will be explained in this chapter. *Seamless mobility* includes handovers to other terminals, to other access networks as well as to other administrative domains. *Generic access* includes authentication and authorization for both network access and service access. *Session control* includes session negotiation as well as session adaptation.

Our approach is based on a federation of distributed software components in the service control layer that run on different hardware (in networks and on terminals) within different administrative domains. The SP federation ensures homogeneous service access across all federated networks, including session mobility across network domains, and provides transparencies to end users and application providers with respect to both the domains and the underlying wireless and mobile network technologies. It enables roaming of end users through arbitrary environments while automatically maintaining connectivity to their Internet TV station, for example, or on-going phone calls. This means that irrespective of the network environment the user is in, the services and applications will always be made available.

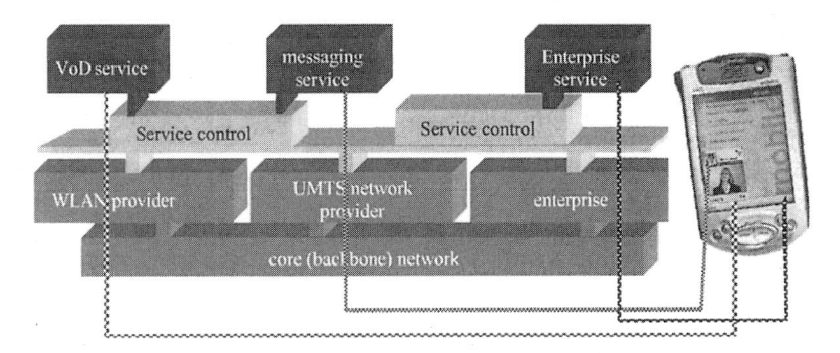

Figure 6-1. Service Platform based on a federated service control architecture [1]

In the current chapter we will explain our SP concepts and show some of our solutions. More detailed results and more results can be accessed via our project website [2] and through our publications [3-8]. The remainder of this chapter is structured as follows. In section 2 we describe the 4G environment and motivation. The concept of federation is explained in section 3, followed by access network selection (in section 4), security and authentication (in section 5), and session control and mobility management (in section 6). Implementation examples are finally presented in section 7, validating our concepts and solutions of access network selection, mobility management, session adaptation and federation. Benefits are briefly described, and finally conclusions are drawn.

2. BACKGROUND AND MOTIVATION

In this section we will illustrate our view on seamless roaming in a 4G environment (see Fig. 6-2) by means of a user scenario and a description of networks and terminals.

2.1 Scenario sketch

Tom, a knowledge worker, is on his way to work, travelling by train. He is equipped with a WLAN and GSM/GPRS capable terminal and he has set up a session via the GSM/GPRS radio interface with premium videoconference service that is offered by an application service provider (ASP). On his way to work Tom passes through several train stations offering IP-connectivity via a WLAN access point. These traveller services offer a cheaper and higher bandwidth service than the public GPRS service (which has a wider coverage area). The WLAN interface of Tom's terminal his videophone call is maintained and handed over - without disruption - between GPRS and WLAN networks. The video quality is adapted to the available bandwidth: with the GPRS connectivity, the video is dropped, leaving only the audio signal. Tom is travelling in rush hour and heavy use is made of the network on the train stations. Therefore at these hot-spots his video call is only presented in a stamp-size format on his terminal. Once Tom arrives at his office premises, his company WLAN network is detected, followed by a log-on and again a seamless session handover of the video service from GSM/GPRS to WLAN. The video is now presented full size with high resolution and refresh rate.

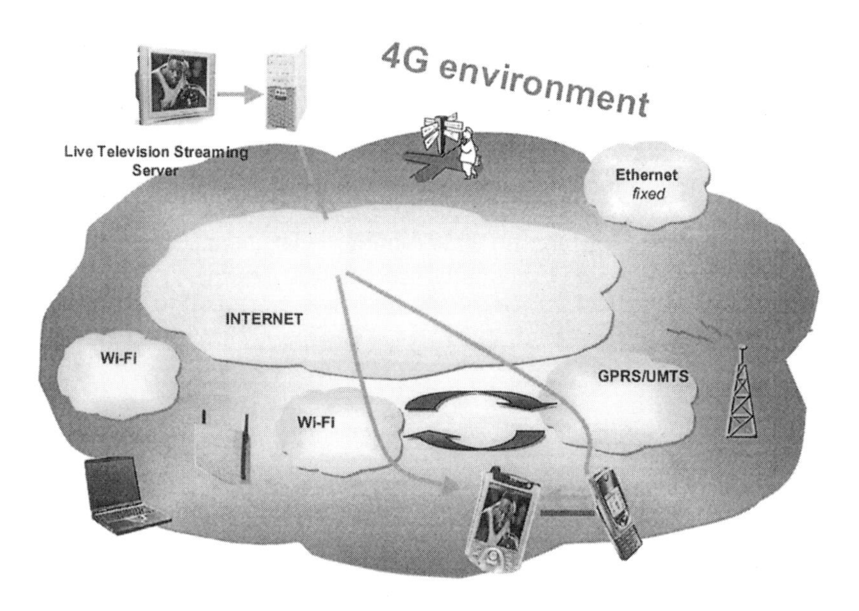

Figure 6-2. Illustration of seamless roaming in a 4G environment [3].

2.2 Network and Terminal description

Mobile users are likely to move from coverage of one access network to another. These access networks can have different characteristics (heterogeneous networks), belong to different administrative domains, offer different features and services with different quality levels (e.g. bandwidth), and cost differently. However, it is widely accepted that there is global IP-connectivity between all networks and all end-to-end communication is IP-based. Heterogeneous access networks are of two generic types: wired (e.g., xDSL and Ethernet) and wireless networks (e.g. UMTS and Wi-Fi™).

based. Heterogeneous access networks are of two generic types: wired (e.g., xDSL and Ethernet) and wireless networks (e.g. UMTS and Wi-Fi™). Wireless access networks can, in turn, be categorized according to their scope and bandwidth into Wireless Wide Area Networks (WWANs) and Wireless Local Area Networks (WLANs). Example WWANs are GPRS and UMTS cellular networks, and example WLANs are Wi-Fi™, HiperLAN/2 and HomeRF networks. Compared to Wi-Fi™, GPRS/UMTS provides a lower bandwidth per unit area and connection over a larger geographic area.

On the client side a number of basic assumptions regarding terminal characteristics are used as starting-points. In our SP architecture, a user's mobile terminal (e.g. PDAs, telephones, laptops) generally is a small but full-fledged open computing system that can run applications from different sources. It is configured with a modern general-purpose operating system (Windows CE, Symbian, Linux, etc.) and a standard IP stack to support the communication with other users and services over the Internet. It has a powerful modern CPU, a relatively big colour screen, and sufficient memory to store and run additionally installed software. Examples of existing products are the new Symbian based mobile terminals from Nokia (3650, 7650) and SonyEricsson (P800), but also devices based on PocketPC and Linux in PDA format. The mobile terminal may be equipped with multiple network interfaces to enable network connectivity using different types of access networks such as 2.5G and 3G mobile networks, WLAN wireless hotspots, but also Ethernet to fixed networks. We do not consider power consumption issues and assume that all wireless interfaces can be active at all times.

3. SERVICE CONTROL

Delivery of services via heterogeneous networks adheres to a 3-layered model consisting of an application layer, a service control layer and a transport layer. The transport layer consists of heterogeneous access networks and core networks. The application layer contains all application logic needed to provide services from Application Service Providers (ASPs) to mobile users. The service control layer is logically located between the application and transport layers and shields the network heterogeneity for the different parties.

Fig. 6-3 shows a schematic structure of the operating environment that includes the components of the three layers mentioned above. The service control layer makes use of service platforms interoperating through federation. The service platform(s) in the control layer offer(s) service control functions that enable end users to easily gain and maintain access to their services to end-users and hides the changes of access networks and terminals due to roaming of end-users. For access networks, a service platform provides control functions for transport outsourcing services. One of our previous publications [4] describes how service platforms and the envisioned federation among them realize the service control functionality.

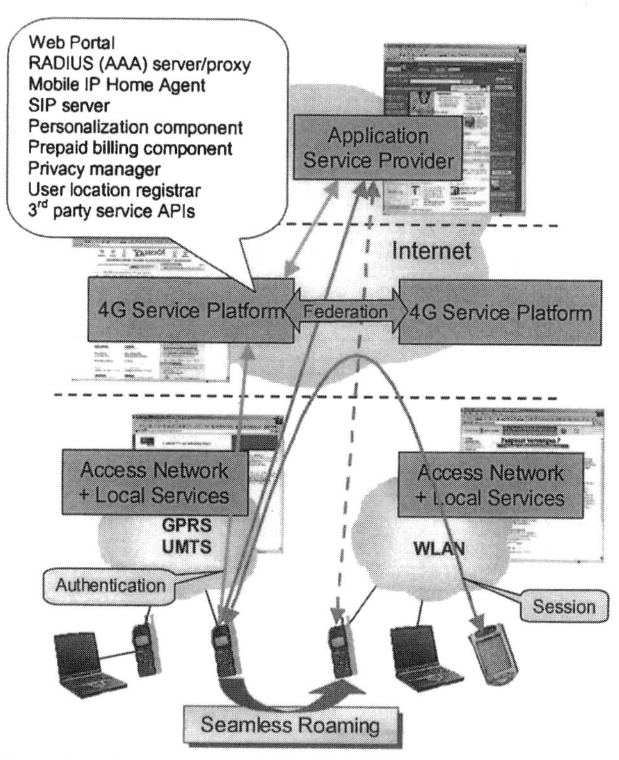

Figure 6-3. Illustration of service platform federation (and the set of functional components in a typical service platform)

The system realizing the service control layer has a distributed architecture that spreads over different entities, including the service platform(s) and the mobile terminals. The federation enables the interaction between the components of peer service platforms. Fig. 6-3 lists a set of typical service control components that reside in a service platform. Fig. 6-4 illustrates the typical service platform components residing in mobile terminals. In the subsequent section some of these functional entities are described in detail.

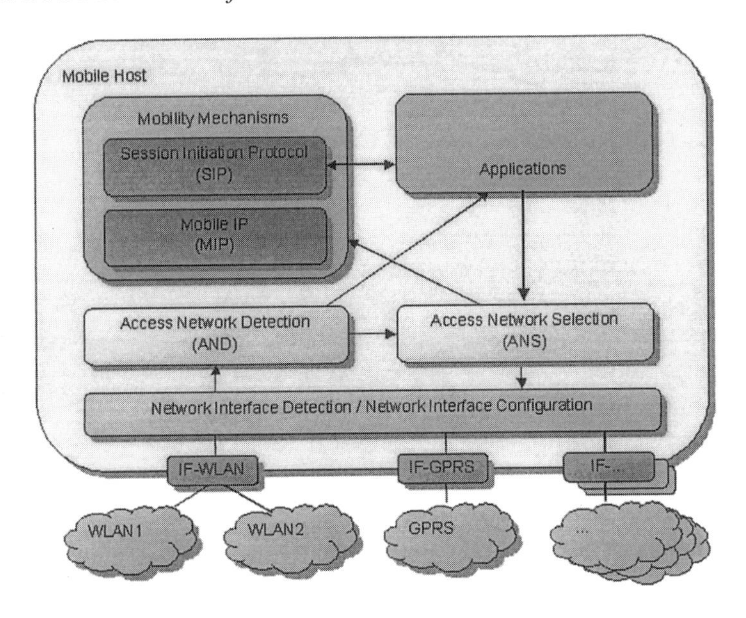

Figure 6-4. Illustration of terminal architecture (only important service control components).

4. ACCESS NETWORK SELECTION

The user entering the train station in our scenario sketch (section 2) has a choice of selecting a new (WLAN) access network. The selection of the Access Network (AN) depends, amongst other, on (see Fig. 6-5):

- Application Service Provider (ASP) requirements,
- Provider policies (e.g. agreements between user, access network provider and Service Platform provider about payment and usage rights),
- Access network characteristics (such as available bandwidth),
- User requirements (e.g. about quality of service, costs or service support),
- User subscriptions (i.e. is the user allowed to use the network, does he have the right certificates and user credentials) and
- Terminal capabilities (such as what kind of hardware interfaces are present and active in the terminal).

Information about all these categories is needed to maintain sessions of the mobile user in a seamless way, optimally tailored to his (new) circumstances. This data is available in profiles stored at the terminal, SP, AN, and the ASP and in policies and mutual agreements, which is input for a decision module, which for example could be (and in our case *is*) part of a mobility manager (see section 7).

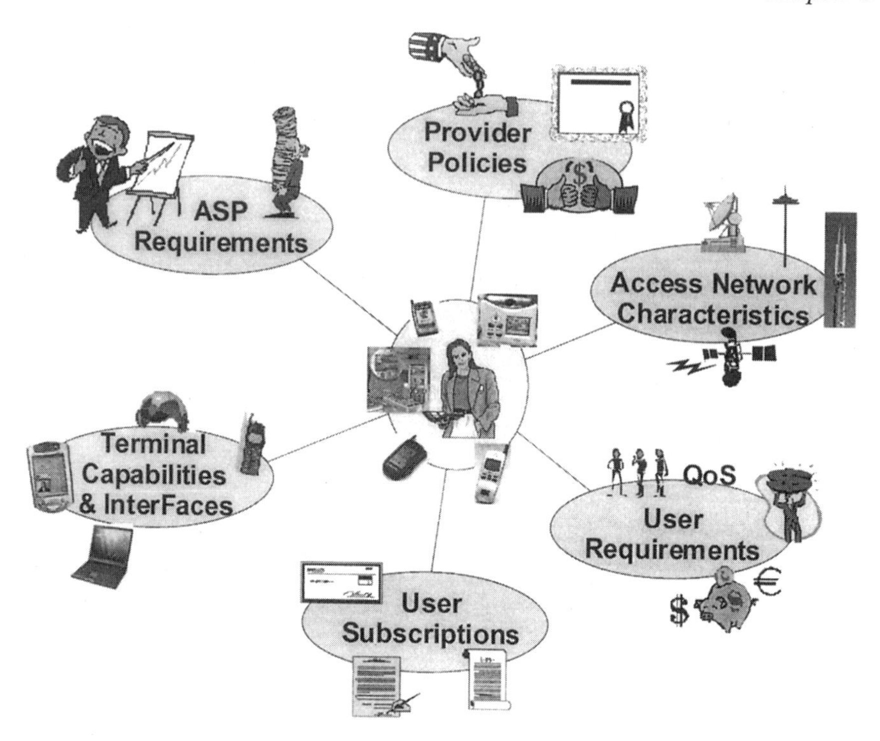

Figure 6-5. Illustration of categories on which Access Network Selection is based on.

4.1 Motivation for Terminal controlled Access Network Selection

In terminal controlled Access Network Selection (ANS) scenario, software running on the terminal determines whether a new Access Network (AN) should be selected. The terminal may need information from the SP in order to make the decision, but the actual decision is made in the terminal. In contrast, in an SP controlled ANS scenario, the SP determines whether the terminal is allowed to switch to another AN. In that case the SP requires information from the terminal about available network and then decides whether the terminal should switch, e.g. based on roaming agreements.

When the ANS decision process runs on the terminal, information about the terminal, its access network environment, its running sessions and its user are much closer at hand. This implies that the terminal typically is in a much better position to make a decision about ANS than the SP. In a 4G environment with many overlaid access networks with different working areas, information about networks is highly dynamic. In terminal controlled ANS this information is directly available at the terminal and can be used for

ANS this information is directly available at the terminal and can be used for ANS. In SP controlled ANS, this information must continuously be communicated with the SP, and moreover the role and control of the user is limited. For one type of network and a clear threshold of signal strength as the basis for a decision this is possible, but in a heterogeneous network environment this is troublesome. Problems can be expected when a user has subscriptions with multiple Service Platform providers, which can lead to contradicting SP ANS policies.

Increased terminal capabilities (processing power, memory, storage, etc) have given network operators and service providers the opportunity to remove some of the load from their network equipment and servers, and distribute it to the terminals. This means that the network and service providers will more and more loose control since advanced terminal capabilities are beyond their control [9]. The trust-requirement of the user to his personal terminal weakens the operators' control even further. The personal terminal is the closest system component to the user and has (stored) knowledge of the whereabouts and contextual information of the user and can be used to identify the mobile user. The role of personal terminals in 4G environments is another drive for terminals to become out of the control of access network operators. A mobile user should be able to trust her/his personal terminal because the terminal, as the closest system component to the mobile user, is in position to reveal to the system the (exact) context information and whereabouts of the user. Thus, the mobile terminal should be able to defend and protect the interests and rights of its mobile user against those of other parties (such as access network operators and service providers). The level to which personal terminals will protect the interests and privacy of mobile users will directly determine the success of service provisioning in next generation environments. Terminals can only be able to fulfil the role of a trustworthy and privacy protecting device as expected by the user if a terminal is also the point of control with respect to selecting networks and connecting to ASP's.

The above-mentioned reasons led us to our choice to base our architecture on terminal controlled ANS.

4.2 The role of the Service Platform on server side

In addition to terminal intelligence for network selection, any client-side application needs SP functionality on the server side. Examples are the SIP functionalities implemented on the server-side that is required for SIP-supported mobility management and session control (section 6), AAA (Authentication, Authorisation, Accounting) functionality (section 5) and profile management. As for the latter: we assume that most of the in profiles. The service platform can provide storage, updating and distribution of profiles. It also provides a portal for the user to set his profiles and privacy requirements. For example, the user can give an ASP permission to use its location for delivering location-based services.

4.3 The role of the mobile user

At some point in the session handover process, access network selection may also require the involvement of the mobile user. This can be the case when the network usage costs rise beyond user permit (as determined in the session settings). Fig. 6-6 shows preferences that users can set and store in a user profile: for example, Tom from our scenario (see section 2) prefers high quality network connections to low quality, but rules out too expensive networks.

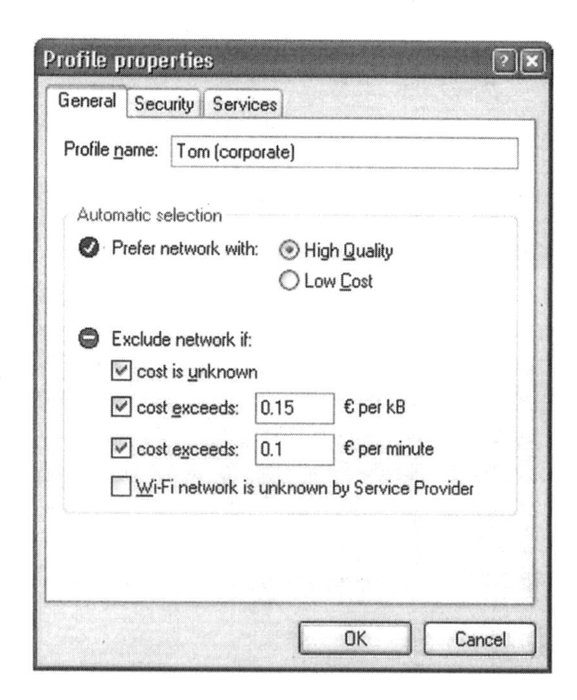

Figure 6-6. Users can store their preferences on e.g. network quality and costs in a profile.

Our design philosophy is "user involvement with a minimum of user interaction". An implementation of this is e.g. to confer the mobile user by means of pop-ups delicately appearing and providing for appropriate information that supports him during the decision process, followed by the means of pop-ups delicately appearing and providing for appropriate information that supports him during the decision process, followed by the provisioning of suitable portals to feed the decision into the system. We gave shape to this by means of pop-ups on network costs for a new available network, see Fig. 6-7.

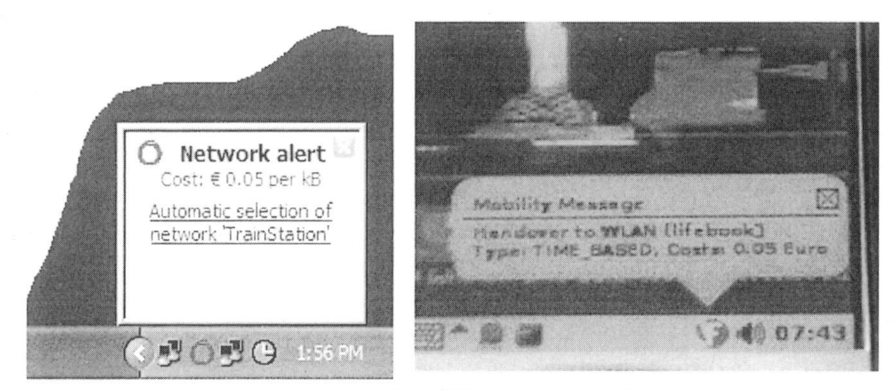

Figure 6-7. Alerts with tariff info on network change.

5. SECURITY AND AUTHENTICATION

Security and authentication are a means to prevent unauthorised use of information and resources, which applies to both network access and service access. Authentication and authorisation should be secure between all involved parties. Not only does the user need to authenticate in order to be authorised for network usage, also the operators must have a trust relationship among each other.

5.1 Federated AAA

In case of WLAN authentication, before any terminal can access a new network, user credentials have to be exchanged between the terminal and an Access Point nearby, followed by an exchange between the home and visited SP domains (federated AAA functionality). Based on the trust relation and agreement between the SP operators the visited SP will allow the terminal access to its network. In the case of SIM authentication (used in GPRS/UMTS), the operator typically has full control (based on an pre-established subscription between the user and the AN operator) and there is no link with the AAA infrastructure of the Service Platform.

5.2 Authentication management

Our starting point is that mobile users have a subscription with an SP. The SP (server side) maintains a database with all subscriptions that include the user's credentials such as his name, how to pay the bill, as well as the public part of the user's client certificate. Certificates enable the unique and secure identification of the user, which can be used for secure access to web pages (e.g. SSL) but also for IEEE 802.1x [10] authentication for WLAN and LAN. IEEE 802.1x is an umbrella for various authentication methods, and the variant using client certificates – either pre-installed or on a smart card or dongle – is EAP-TLS. The advantage of some of the EAP authentication methods, including EAP-TLS, is that no user interaction is required when authenticating for network access, thus allowing the user to move around and roam from one WLAN to another [5].

When a mobile user attempts to get access to a WLAN network or hotspot, the SP with which the user has a subscription may not necessarily own the WLAN network. For almost all network access authentication mechanisms, including 802.1x, the user (via his terminal) must provide his user ID, commonly in the form of name@domain. The domain name can be used to resolve the SP, through a lookup table in a database, DNS or UDDI. Between the WLAN access point and the SP, the RADIUS protocol [11] can be used for authentication. As a result, it is the SP that authenticates the user and not the WLAN operator. Such a scheme is feasible since RADIUS is widely deployed and supports all variants of 802.1x authentication.

In case the WLAN operator provides free access to all users, the WLAN access point may signal the user's terminal that a client certificate is not required for authentication [12]. This is supported by the 802.1x protocol and requires no change on the terminal. However, even with free access, there is an advantage in always forwarding authentication requests to the SP, since the SP will know the network location and possibly geographical location of the end-user. The latter may be important due to legal regulations, thus refraining the WLAN operator for implementing such functionality.

Federation between operators and sharing AAA (forwarding user credentials) makes it possible that the user can authenticate on various access networks. The authentication procedure for getting access to WLAN should be possible without intervention from the user to allow seamless roaming. The user does not need to log on to the SP as well when authenticated for WLAN access. Federated AAA is only possible when trust relationships and SLAs exist.

5.3 Service authentication

Application Service Providers offering web-based services can leverage mobile user information available at the SP. For instance, the SP may offer network location, geographical location and terminal type information, identity services and session-based services (e.g. through the Session Initiation Protocol, SIP) to third parties. For a third party to access such information, there must be a trust relation between the SP and the third party. In addition, the user must have given his consent that a third party uses privacy sensitive information such as his location. With these conditions in place, the end-user accesses the third party's services (e.g. web pages) in a secure fashion. It is possible to reuse the client certificates, where in fact user authentication is 'outsourced' to the SP. As an example, by means of secure web access (e.g. SSL) using the client certificate, the third party can verify the user's credentials with the SP through protocols such as SAML [13] (Security Assertion Mark-up Language). This protocol, which has its origin in the Liberty Alliance initiative [14], is also suitable for returning the types of information that the third party may access.

6. SESSION CONTROL AND MOBILITY MANAGEMENT

A session, as an instantiation of a service, is controlled by Session Control functionality that cooperates with Mobility Management functionality. Session control includes a set of non-service specific procedures to establish, maintain and terminate sessions and includes interactions between controlling parties and resources. The Mobility Management functionality is in charge of locating mobile users for session initiation and handing over their sessions between different network attachment points, as users roam across terminals and access networks. This section provides a high-level description of these functions and how they interact.

6.1 Session control

The Session Control function of the 4GPLUS Service Platform takes care of the creation, maintenance and termination of a session. Session Control comprises a set of generic sub-functions carried out in the so-called session initiation, session maintenance and session termination phases of a session lifecycle.

Session initiation phase includes the following generic functions: naming of a mobile user (the session initiator finds out the static, i.e. location independent, address of the mobile user's terminal), locating of the end-user (the session initiator finds out the current address of the mobile user's terminal), inviting of the mobile user (the session initiator sends an invitation to the mobile user), negotiating about session requirements and settings (the end-parties involved in the session implicitly or explicitly decide on the session settings), and reserving resources needed for establishing the session according to its settings.

Session maintenance phase includes the following generic function: monitoring of a session content by authorized parties (e.g., for lawful interception and administration purposes), and modifying session settings during the lifecycle of the session (e.g., by adding/removing new media or inviting other users). A session modification can be triggered manually by the end-user or automatically by a system component. The Mobility Management component may trigger the Session Control component when the session route is modified due to user or terminal movement.

The session termination phase encompasses the generic function of tearing-down of existing sessions. When a session is broken down, this function releases all corresponding resources for future sessions. Session termination may originate with a user request, remote disconnection by the other party, or for application-specific reasons.

6.2 Mobility management

Mobility Management is a functional component of the Service Platform that firstly keeps track of the IP-addresses of mobile users, and secondly modifies the IP routes of the ongoing sessions of mobile users. These high level functions of Mobility Management are referred to as Location Management and Handover Management, respectively [15].

Location Management enables the Service Platform to discover the current attachment point(s) of a mobile user that will enable other users to initiate new sessions towards the mobile user. The term "location" here refers to the IP address(es) of the terminal interface(s) to which the mobile user is connected. Other users willing to initiate a new session towards the mobile user contact the Location Management unit that either returns the current IP address of the mobile user to the session initiators or forwards the session initiation request towards the mobile user's current IP address. Location Management encompasses the following functions: updating of current terminal addresses (i.e., address registration), handling of session invitations (triggered in the session initiation phase to find the whereabouts of mobile users' terminals), and Paging (to find the exact whereabouts of an idle mobile terminal associated with a mobile user).

Handover Management function enables the Service Platform to maintain a mobile user's *ongoing session(s)* as her/his corresponding IP-address(es) changes(change). The need for a session handover is triggered by, for example, detecting user and/or terminal movement with respect to an access network, changing network conditions and load, changing session usage cost, etcetera. The sessions of a mobile user can be handed over collectively or separately. The functions carried out within the realm of the Handover Management component can be grouped in two generic functions: regenerating of a session route (to find and get the resources for the new session path), and controlling data flow (to maintain the delivery of the data from the old session path to the new session path, according to the session requirements) [6].

6.3 Inter-working

An overlap between the functionalities of Mobility Management and Session Control exist in establishing and controlling sessions in heterogeneous and dynamic environments. This overlap implies that there should be a close interaction between these SP components and that the mobility management component should be session-aware [7,8].

A single existing handover technology is not likely to be optimally suited for handing over the complete session context under all possible conditions. Mobile Internet Protocol (Mobile IP) [16] and Session Initiation Protocol (SIP) [17] are commonly known solutions in this area (and are also used by us). As a network layer solution, Mobile IP is suitable for 'mobility unaware' applications with relatively long-lived connections or the requirement to have a constant IP address. Examples are applications for downloading, shared file access and messaging. SIP on the other hand is suitable for 'mobility aware' applications. For instance, multimedia applications may re-register with the SIP server and transfer active sessions during network handover, i.e. performing mobility management. The latter allows for continuation of active sessions using another network, similarly to Mobile IP. However, sessions can also be adapted to suit the characteristics of the new network in an optimal fashion, i.e. performing session control. For instance, when switching from a WLAN network to a cellular GPRS network, the SIP application may use lower quality video or drop video altogether, leaving an audio session only.

To support both mobility unaware and aware applications we use a combination of Mobile IP and SIP. This requires a Mobility Manager at the mobile terminal that manages the network handovers by controlling the

Mobile IP client software and notifying SIP applications of network handovers. In the 4GPLUS project [2], we have implemented Mobility Managers on the Windows and Linux platform and realized Mobile IP and SIP inter-working.

7. VALIDATION EXAMPLES

Within our 4GPLUS project [2] we realized several validators. Detailed descriptions of the concepts [4-8] and implementation work are published, e.g. [7,8]. Here we show a couple of examples illustrating the research described in the previous sections. Our proof-of-concept work is carried out on two different terminals, being a PDA running Linux (section 7.1) and a laptop with Windows XP (sections 7.2 and 7.3).

7.1 Validation example of Access Network Selection, Mobility Management and Session Adaptation

In an environment with multiple available wireless packet-switched networks, a mobile host may have the opportunity to obtain IP connectivity using each of these access networks. At any time, the mobile host may have none, one, or multiple network interfaces active to provide connectivity to the applications it runs. Additionally, the active network interfaces can be associated with several but different kinds of access network technologies, e.g. based on UMTS, WLAN and Bluetooth. The mobile host chooses to activate one of the corresponding network interfaces depending on, for instance, the access network availability, the cost of network usage, the link layer quality, and/or the preferences of the user. Even if the TCP and UDP connections are maintained at system level, where handover from one access network to another is managed by a technology such as Mobile IP, this means that the applications running on the mobile host experience changes in available network resources and in network characteristics. Some types of mobile applications may be interested to receive information about the current resource status, i.e. may be able to anticipate changes in the set of active network interfaces, changes in the access network used for existing connections, and therefore changes in the available network bandwidth. For instance, a multimedia player may benefit from knowing upfront that a default access network change is going to take place, and may alter the used codecs on the fly to better suit the expected network characteristics.

Our research results and detailed information describing host mobility management is published [7]. It includes a mechanism that provides status information about the current network resource and roaming situation information about the current network resource and roaming situation towards the applications running on a mobile host. The mechanism builds on top of the *mobility management* functionality at the mobile host that is responsible for detecting and activating the access networks and for interacting with the Mobile IP functionality.

The validation is done with a streaming-TV application [7]. The video streaming server is a Linux host running a VIdeo Conference tool (VIC) configured in transmit-only mode, thus used only for sending out a video stream. The video stream is a RTP unicast UDP stream of an encoded, captured TV signal. The video streaming client running on the mobile terminal is also a VIdeo Conferencing tool (VIC) that is configured in receive-only mode. The VIC client receives the RTP packets, decodes the h.261 frames and displays the video stream in a local window, see Fig. 6-8 Extended functionality is added to the VIC server and VIC client in order to enable a peer client to dynamically reconfigure the server and adapt the running streaming session characteristics (comparable to the usage of a SIP re-invite transaction). This adaptation-control decision is typically done upon interpreting the messages from the mobility manager about any changes in the network environment (AND and ANS, see also Fig. 6-4) and the state of Mobile IP. Examples are given in the screenshots in Fig. 6-8 where the presented video image on the PDA is stamp-sized at the low bandwidth case and full screen sized at high bandwidth.

Details on the access network detection and selection process in the mobile host are as follows. First, a detection mechanism scans for new networks and monitors the quality of active networks. The detection process results in a list of available networks, their names (network or operator name) and their signal strengths, see Fig. 6-9. This list may contain a priority order for the networks. Second, an authentication infrastructure is built and running at the server side of the SP. This AAA infrastructure supports both the terminal and the AN to perform AAA tasks and enables client devices to check the accessibility of the newly discovered networks. Based on up to date information about the AN, the user requirements and subscriptions, the application or service requirements and provider policies, the decision process decides whether the currently active AN can still be used for the current session or whether another AN has to be selected from the list and activated. The user is provided with information about the network by pop-ups (compare section 4.3). Such information is requested from the SP instead of the network itself. The advantage of this approach is that the terminal obtains information about network usage costs and quality of the network *before* connecting to it. The user can also do ANS by manually selecting one of the available networks, thereby overriding the automatic selection process.

Figure 6-8. TV signal received at the PDA with Ethernet (left figure) and GPRS (right figure) connectivity showing session adaptation.

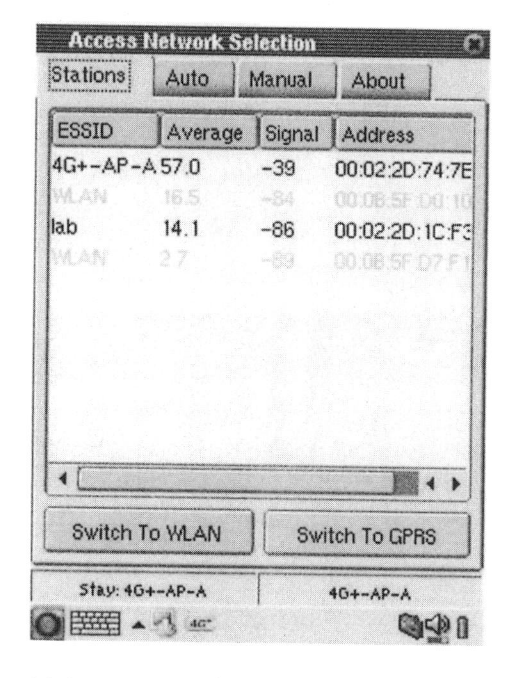

Figure 6-9. Screen shot of PDA during access network selection [1].

7.2 Validation example of federation of AAA

We have realized various WLAN hotspots supporting authentication with multiple IEEE 802.1x variants (such as EAP-TLS, EAP-TTLS and EAP-SIM) and RADIUS authentication to the SP. We have even connected our WLAN hotspots and SP platforms to the national Dutch SURFnet infrastructure [18]. As a result, 4GPLUS users and users from various universities, research institutes and companies have access to hundreds of WLAN hotspots throughout the country.

In addition to network authentication, we have realized federation of SP platforms. By integrating our service authentication components in the WASP service platform [19], both 4GPLUS and WASP users can securely access web-based and location-aware services offered by the WASP project. Similarly, these users can securely access a hotspot location service, which shows WLAN hotspots near the user. This service (of which the schematics are shown in Fig. 6-10) has been developed by 4GPLUS using WASP service functionality.

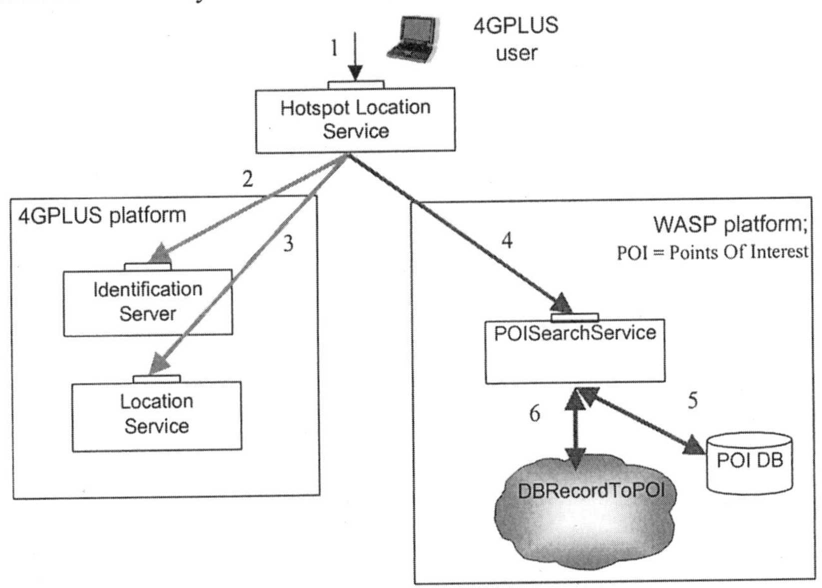

Figure 6-10. Schematics of federation validation

7.3 Validation example of inter-working

In 4GPLUS we have developed the MX (Multiple Access) SIP Communicator based on an open source Java implementation of a SIP client screenshot of an ongoing session is shown in Fig. 6-11. The MX SIP Communicator works together with the Mobility Manager to support mobility management and session control in a synchronised fashion. Both Mobile IP and SIP can be used for mobility management to support seamless roaming of the user from one network to another. SIP handles session control by adapting the session based on the network characteristics (e.g. use of audio and video on LAN networks and audio only on GPRS networks).

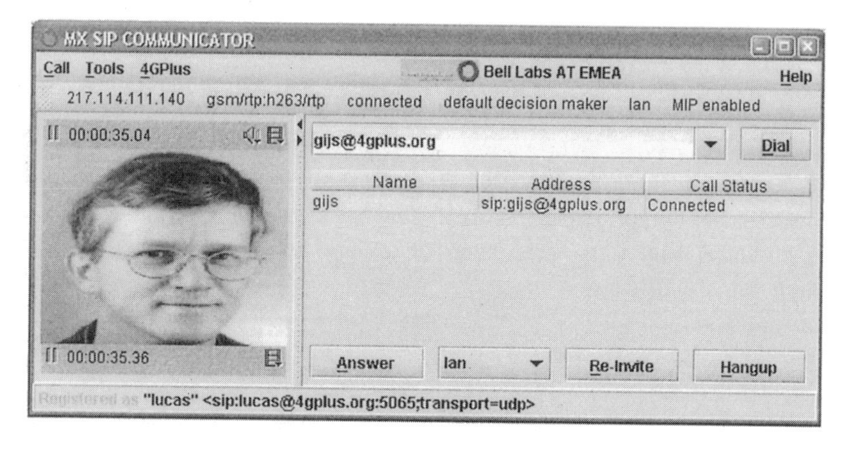

Figure 6-11. Screenshot of the MX SIP Communicator with an ongoing audio/video session.

We have demonstrated that both Mobile IP and SIP can be used for mobility management. However, since the MX SIP Communicator is a mobility aware application, we have also demonstrated the adaptation of multimedia sessions [8]. For instance, when switching from a fixed LAN to a GPRS connection (triggered by removing the LAN cable), the SIP application automatically drops the video stream but leaves the audio stream intact. Similarly, when switching back to LAN, video is automatically added again. This works both with and without Mobile IP due to the inter-working with the Mobility Manager for synchronization.

Session control with SIP only shows better performance than SIP over Mobile IP during network handovers in the order of 1 (LAN and WLAN) to 4 (GPRS) seconds. SIP over Mobile IP (using co-located care-off-address and UDP tunnelling) shows minimum handover times of 2.5 seconds and maximum values of 14 seconds, which can be explained by the fact that SIP control messages over UDP are retransmitted with a logarithmic back-off time in case of late acknowledgments.

8. BENEFITS

The approach of our service platform incorporates existing business and parties. Moreover, it introduces new business opportunities, by offering complicated or costly platform functionalities to (small) enterprises. The latter can use e.g. database profiling, authentication, and handover management functionality from the SP.

As an example consider small access network operators. These are not likely to have subscriber databases for authentication purposes for all of the visiting users. This is especially true for privately owned WLANs, for instance someone living near a bus stop who would like to provide Internet access to anyone, hoping to get some money for it. Assuming that a user has a subscription with an SP, only the SP knows the user and can therefore handle the authentication and authorisation for the WLAN. Once a user is authenticated, the AN provides IP connectivity to the user.

In addition, our SP approach of hiding the heterogeneity of the 4G environment (different access networks, terminals, user preferences, services) is beneficial for all concerned: end-users, service providers, access network operators. Especially seen from the mobile user perspective the advantages are:

- Automatic selection of access network (based on preferences),
- Single (zero) sign-on,
- Single subscription with one 4G Service Platform,
- Always and everywhere connected and reachable, independent of location,
- Same services and service context independent of the location of the user,
- Enhanced service offering such as location based services,
- Privacy and profile management where the end-user is in control, and
- Maintain session during hand-over.

Since terminals normally are considered as personal devices, having client-side SP functionality is "handy": users e.g. always carry their credentials and preferences with them. Besides it enlarges trust of users; their privacy sensitive information is stored in their personal terminals that only they have access to. Business is therefore likely to profit from this.

9. CONCLUSIONS

Technological solutions enabling seamless roaming in heterogeneous environments are researched and validated, basically by using federated service platform (SP) functionalities as a starting point. Through co-operation of the distributed SP components, the interoperability issues between the various (network) technologies and administrative domains are solved. Future terminals will be equipped with multiple network interfaces to enable network connectivity using different types of access networks. The end users will at some point move out of coverage of a particular access network. Seamless mobility, i.e., the capability of end users to roam through heterogeneous environments while automatically maintaining connectivity to their application, is realized by federated mobility management functions. Various kinds of access control, mobility management and inter-domain AAA are put in place to make this work. Also session adaptation to the changing environment is realized, as are appealing end-user services. Important key protocols used in our solutions include Mobile IP, SIP, SAML and IEEE 802.1x with different EAP variants.

Both Access Network Detection and Access Network Selection functionality have been implemented in a Mobility Manager running on the terminal. Validation examples included our developed host mobility manager and the MX SIP Communicator. These implementations show how applications running on the terminal can be made aware of the characteristics of the available network resources and how they can respond to that. Furthermore they show how different mobility protocols (e.g. SIP and mobile IP) can be combined into one infrastructure. By making use of context awareness, the roaming experience is transparent for users: seamless handovers are triggered and implemented on the client-side without requiring interaction with the user changes of access network and network capabilities. The automatic selection functions use the access network characteristics (availability, signal strength, costs) as well as the user context (location, user preferences, subscription information), terminal capabilities and session context.

More research has to be done on the scalability of our solutions, for example dealing with peak loads when a train arrives at a station and many passengers will try to access the station's WLAN service at the same time. In order to be able to manage the network load, information about the current load of an access network should be communicated to the SP. Communicating such dynamic network load information between AN and SP is straightforward in our solution. Thus, our architecture and our implementations on both server-side and client-side make it possible to continue multimedia sessions in all circumstances.

ACKNOWLEDGEMENTS

Elements of the work described in this chapter are part of the research project called 4GPLUS (4th Generation Platform Launching Ubiquitous

Services), see http://4GPLUS.Freeband.nl. This project is supported in part by the Dutch Freeband Impulse Programme on Tele-communication Applications (www.freeband.nl) and is co-funded by the Dutch Ministry of Economic Affairs.

REFERENCES

1. H. van Kranenburg, R. van Eijk, M.S. Bargh, A.J.H. Peddemors, H. Zandbelt, and J. Brok, "Federated Service Platform Solutions for Heterogeneous Wireless Networks," *Proceedings DSPCS'2003* (7th Int. Symp. on Digital Signal Processing and Communication Systems), Australia , 8-11 Dec. 2003.
2. 4GPLUS (4th Generation Platform Launching Ubiquitous Services) project; http://4gplus.freeband.nl.
3. H. van Kranenburg (also ed.), J. Koolwaaij, J. Brok, B. Vermeulen, R.L. Lagendijk, J. van der Meer, P. Albeda, D-J. Plas, B. Busropan, and H. Eertink, "Ambient Service Infrastructures - Supporting tailored mobile services anytime, anywhere," *Freeband Essentials 1;* available at http://www.freeband.nl, Jan. 2004.
4. M.S. Bargh, J.H. Laarhuis, and D-J. Plas, "A Structured Framework for Federation between 4G-Service Platforms," *PIMRC 2003*, Beijing, China, Sept.2003.
5. Hong Chen, M. Živković, and D-J. Plas, "Transparent End-User Authentication Across Heterogeneous Wireless Networks," *Proceedings of IEEE VTC 2003 Fall*, Orlando, FL, USA, 4-9 Oct., 2003.
6. M.S. Bargh, H. Zandbelt, and A.J.H. Peddemors, "Managing Mobility in Beyond 3G-Environments," *Proceedings of the 7th IEEE international conference on High Speed Networks and Multimedia Communications* (*HSNMC 2004*), Toulouse, France, June 30-July 2, 2004,
7. A.J.H. Peddemors, H. Zandbelt, and M.S. Bargh, "A Mechanism for Host Mobility Management supporting Application Awareness," *MobiSys2004*, Boston, Massachusetts, USA, June 6-9, 2004.
8. W.A. Romijn, D-J. Plas, D. Bijwaard, E. Meeuwissen, and G. van Ooijen, "Mobility Management for SIP sessions in a Heterogeneous Network Environment," *Bell Labs Technical Journal*, BLTJ **9**(3), 2004.
9. D. Chantrain, K. Handekyn, and H. Vanderstraeten, "The soft terminal: service intelligence from the network to the terminal," *Alcatel Telecommun. R.* **2**, 135-141 2000.
10. IEEE 802.1X-2001. "IEEE Standards for Local and Metropolitan Area Networks: Port-Based Network Access Control," http://standards.ieee.org/reading/ieee /std/lanman/restricted/802.1X-2001.pdf
11. IETF, RFC 2138, "Remote Authentication Dial in User Service (RADIUS)", April 1997; http://www.ietf.org/rfc/rfc2138.txt
12. IETF, RFC 2977. "Mobile IP Authentication, Authorisation, and Accounting Requirements", Oct. 2002; http://www.ietf.org/rfc/rfc2977.txt
13. SAML; http://www.oasis-open.org/specs/index.php#samlv1.1
14. Liberty Alliance initiative; http://www.projectliberty.org/
15. I.F. Akyildiz, J. McNair, J.S.M. Ho, H. Uzunalioğlu, and W. Wang, "Mobility management in current and future Communication Networks," *IEEE Network Magazine* **12**(4), 39-49 (July-August 1998).

16. Mobile IPv4, ftp://ftp.rfc-editor.org/in-notes/rfc3344.txt
17. IETF, RFC 3261. SIP: "Session Initiation Protocol," June 2002; http://www.ietf.org/rfc/rfc3261.txt
18. SURFnet; http://www.surfnet.nl
19. WASP (Web Architectures for Service Platforms) project; http://wasp.freeband.nl
20. NIST SIP Communicator; https://sip-communicator.dev.java.net/

Chapter 7

REESTABLISHMENT OF HEADER COMPRESSION STATE BY CONTEXT TRANSFER IN MOBILE IP NETWORKS

Ha Duong, Arek Dadej and Steven Gordon
Institute for Telecommunications Research, University of South Australia, Mawson Lakes Blvd, Mawson Lakes, SA 5095, Australia

Abstract: Header Compression is a useful technique for reducing the load on bandwidth-scarce wireless links. Header Compression depends on the establishment and synchronization of context at the compressor and decompressor. In mobile wireless networks, it is desirable to transfer this context between access routers to avoid the expensive process of context re-establishment when handovers are required. In this chapter, we propose a method for avoiding the need for context re-establishment when packets are lost during handovers, as well as a method for efficient transfer of context to the new access router (by piggybacking the context information on Mobile IP signaling messages). The analysis of our scheme shows that it can reduce signaling load during handovers, which is beneficial in mobile networks with frequent handovers (e.g. wireless LAN hot spots).

Key words: Header Compression; Context Transfer; mobility; WLAN; mobile IP.

1. INTRODUCTION

Mobile wireless networks offer users the flexibility of mobility, but the network performance may often be compromised due to the bandwidth limitations of wireless link. Therefore, it is desirable to utilize the wireless link bandwidth efficiently. Header Compression (HC) is considered an effective method to reduce the large IP header overhead when transmitting voice packets over the wireless link. At the expense of extra processing at the transmitting and receiving nodes, HC can reduce the load on the wireless link by 50-70% [3].

In general, the HC process is initialized by establishing the HC state at the sender's compressor and the receiver's decompressor. This HC state, or HC context, is continuously updated at both compressor and decompressor upon receipt of each new packet. In mobile wireless networks, as packets are lost during handovers, so will be the HC context, and consequently, the Mobile Node (MN) and the new Access Router (AR) have to re-establish the context. As context re-establishment can be a time-consuming process and requires additional bandwidth, transferring the HC context from the old AR to the new AR, either via the fixed network or via the MN, is a method preferred over context re-establishment [5]. However, even with HC context transfer, if packets are lost during handover, the compressor and decompressor can become miss-synchronized, forcing a full context re-establishment. This problem illustrates one of the difficulties in transferring context that may change dynamically.

Context transfer has emerged as a one of the problems that must be solved in order to achieve full user mobility in mobile wireless networks, with others being routing (essentially solved by the Mobile IP), security and Quality of Service (QoS). Generally, context is the information on the current state of a service provided by an AR. Currently, HC, QoS, security, Authentication, Authorization, and Accounting (AAA) have been identified as candidate services for context transfer [4]. The Koodli and Perkins' work [9] was one of first that mentioned context transfer in mobile wireless networks. The authors, suggesting that context transfer be a part of fast handover signaling, demonstrated the feasibility of fast network layer handover and HC context transfer. However, their approach still requires explicit signaling between ARs and does not take into account the possibility of miss-synchronization between compressor and decompressor caused by packet loss during handover period. Several Internet drafts on context transfer protocols are available, e.g. [6] and [8], suggesting protocol requirements or proposing generic protocols, however these have not been evaluated through analytical, simulation or test bed means. It is also unclear how these protocols would deal with dynamic context such as the HC context.

In this chapter, we present a simple and effective approach to dealing with the miss-synchronization of the HC context by using three-state HC schemes. We also suggest a method of transferring the context, based on Mobile IP signaling. The rest of the chapter is structured as follows. In the next section, we provide an overview of HC schemes; an analysis of HC context establishment cost and a brief description of Mobile IP registration process, and describe the miss-synchronization problem in more detail. In sections 3 and 4, we describe our proposed solution and present simulation results to illustrate its effectiveness. Finally, we give some concluding remarks, and comment on areas intended for future work.

2. BACKGROUND

2.1 Overview of Header Compression

A number of header compression schemes have been proposed for existing Internet protocols, such as in [1], [7], and [10]. In some cases, for example when using IPv6/UDP/RTP, headers can be reduced from 84 bytes to 1 byte, and therefore the header compression scheme can bring up to 73% traffic reduction for a voice packet with 30 bytes payload. The basic principle of the HC is based on the fact that there is a significant amount of redundant information transmitted in the headers of packets exchanged during a session. In other words, most of the header field values remain the same over the lifetime of a session. For a non-TCP session, almost all fields are constant, whereas for a TCP session, there are several constant fields while others change in a predictable way. The header fields can be classified based on their behavior during the session lifetime. For example, the HC scheme in [7] classifies the header fields into:

NOCHANGE (STATIC): the field is not expected to change.

DELTA: the field may change often but usually by a small value (delta) so that the delta value can be sent instead of the absolute value of the field.

RANDOM: the field is expected to be random and thus needs to be transmitted on the "as is" basis.

INFERRED: the field value can be inferred from other values.

Example IPv4 header field classification is presented in the Fig. 7-1. Majority of the IPv4 header fields are STATIC, except for the Identification field which is DELTA, and the Total Length and Header Checksum that are considered INFERRED.

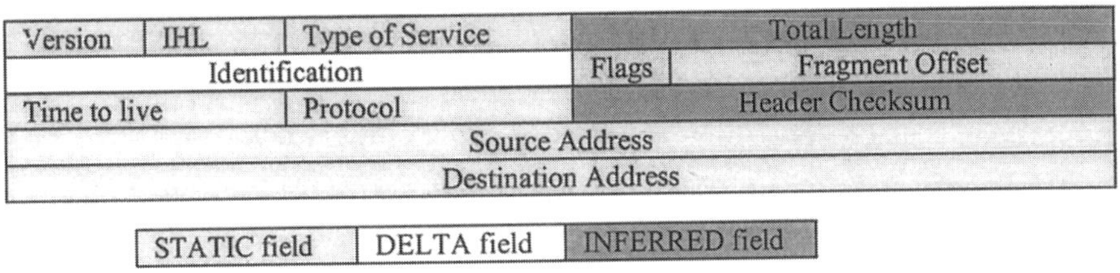

Version	IHL	Type of Service	Total Length	
Identification			Flags	Fragment Offset
Time to live		Protocol	Header Checksum	
Source Address				
Destination Address				

STATIC field	DELTA field	INFERRED field

Figure 7-1. A Classification of the IPv4 Header Fields.

The compression process is initialized with the establishment of the HC context at both compressor and decompressor. The compressor examines packet headers, copies the values of packet header fields to establish the HC context for the new connection, assigns a context identifier (CID) to the established context, and then transmits the full packet (uncompressed), with CID, to the decompressor. The CID can be transmitted, for example, in the Total Length field of the IPv4 packet header. Upon reception of a full header packet, the decompressor compares the actual value of the payload length with the value stored in the Total Length field. The difference between the two values serves as a signal to notify the decompressor about a new HC session. Then, the decompressor starts to build the HC context using packet header fields. The HC context consists of the STATIC and DELTA fields. Once the HC context is established, packet headers can be compressed by not including the STATIC fields, and by using fewer bits to store delta values. Upon receiving a compressed packet header, the decompressor restores the full header by using delta values (if present) and the stored HC context identified by the CID.

To increase the probability that the decompressor correctly establishes the HC context, $\boldsymbol{m}$ full header packets ($\boldsymbol{m>1}$), with CID, are sent initially as part of context establishment. The decompressor is required to receive at least $\boldsymbol{n}$ ($\boldsymbol{n < m}$) full header packets so that it can be confident about the new HC session. The decompressor may optionally send an *Acknowledgment* (*Ack*) message to confirm the establishment of the HC context. To further enhance reliability of compression, some HC schemes have suggested a multi-state HC context. For instance, the scheme in [1] employees a three-state HC context as illustrated in Fig. 7-2.

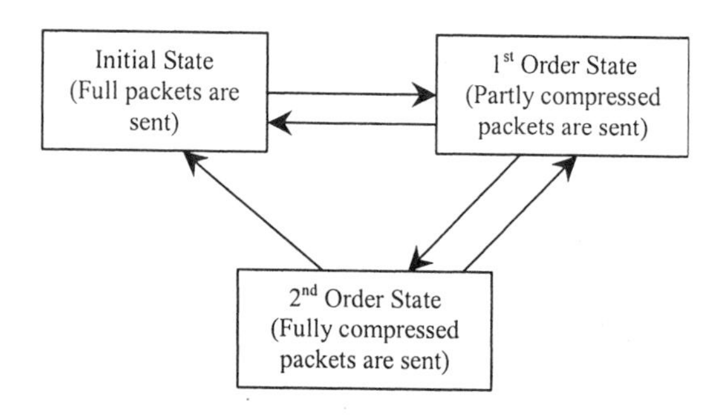

Figure 7-2. Three state Header Compression scheme

In the partly compressed packet, the packet header consists of CID, RANDOM fields, DELTA fields and possibly INFERRED fields. Therefore, the STATIC fields will make up the HC context. In the fully compressed packet, the packet header consists of CID, RANDOM fields, and delta values. Thus, the STATIC and DELTA fields are stored as the HC context, and the DELTA fields are updated whenever a compressible packet arrives. This three-state scheme enhances performance by limiting the impact of lost HC context when in the first order (partly compressed) state.

2.2 Cost of Header Compression Context Establishment

In this subsection, we will analyze establishment cost of HC context (expressed in number of bytes), to motivate the approach of HC context transfer. In practical implementations, the HC context is built in two modes, namely, optimal and acknowledged, depending on channel conditions or, more specifically, the probability of packet loss p_{loss} in the channel. Recall from the previous subsection that the HC context can only be established if the decompressor receives n full header packets with CID. To simplify the discussion, we assume a two-state HC scheme, initial state and compressed state.

In the optimal mode, usually applied when p_{loss} is very low, the compressor sends m full header packets with CID and hopes that the decompressor can receive at least n ($n < m$) full header packets. Fig. 7-3 illustrates the optimal mode of HC context establishment with $m = 4$, $n = 3$. If the establishment is successful i.e. the decompressor receives the required number n of full header packets (Fig. 7-3 a), the cost will be the difference between full header size (f) and compressed header size (c) in m packets. If the establishment fails (Fig. 7-3 b), the extra cost will be a number of compressed packets that have to be discarded ($n_{discarded}$) until the compressor receives a notice from the decompressor, and the number of notices (n_{notice}) sent by the decompressor. Therefore, the establishment cost can be expressed as follow

$$C_{optimum} = \begin{array}{l} m(f - c)P(n \mid m) + \\ \left[m(f - c) + (c + d)n_{discared} + s_{notice}n_{notice}\right](1 - P(n \mid m)) \end{array} \quad (7.1)$$

where d is data (payload) size of the packet, s_{notice} is the size of a notice. $P(n|m)$, the probability of receiving n out of m packets sent, can be calculated as shown below:

$$P(m \mid n) = \sum_{j=n}^{m} \binom{j}{m} p_{loss}^{m-j} (1 - p_{loss})^{j} \quad (7.2)$$

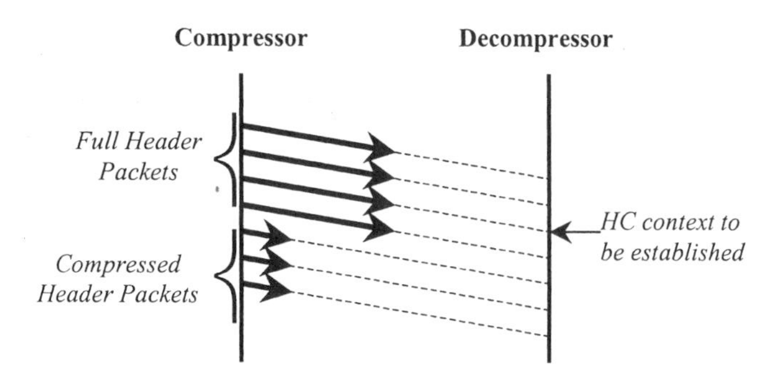

a) A successful case of HC context establishment

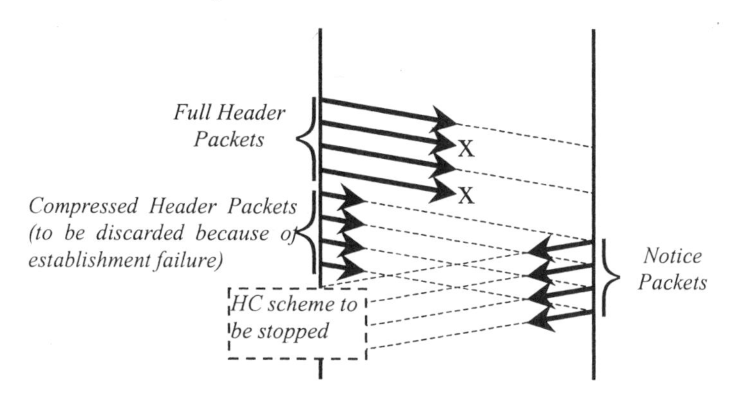

b) A failure case of HC context establishment

Figure 7-3. HC context establishment in the optimal mode with ($\boldsymbol{m}$ =4, $\boldsymbol{n}$=3).

The number of compressed packets discarded by the decompressor depends on the round trip time (***RTT***) between the compressor and decompressor, packet rate ($\boldsymbol{r_{pk}}$) and the probability of the compressor receiving the notice about establishment failure. The dependence on ***RTT*** in the case of establishment failure can be detailed as follows. After sending ***m*** full header packets, the compressor starts to send compressed packets. At the time when the decompressor receives the first compressed packet and sends the notice, the compressor has already sent ($\boldsymbol{RTT/2}$ x $\boldsymbol{r_{pk}}$) compressed packets; they will be discarded at the decompressor. As it takes the amount of time equal to ***RTT/2*** for the notice to reach the compressor, there are another ($\boldsymbol{RTT/2}$ x $\boldsymbol{r_{pk}}$) compressed packets sent that will be discarded. If we assume that notices also suffer from the probability of loss, every lost notice will result in one more discarded packet. In summary, we can express the number of discarded packets and the number of notices as follows

$$
\begin{aligned}
n_{discarded} &= (r_{pk}RTT)(1-p_{loss}) + (r_{pk}RTT+1)(1-p_{loss})p_{loss} + \ldots \\
&+ (r_{pk}RTT+k)(1-p_{loss})p_{loss}^{k} + \ldots
\end{aligned}
\tag{7.3}
$$

$$
n_{notice} = (1-p_{loss}) + 2(1-p_{loss})p_{loss} + \ldots + k(1-p_{loss})p_{loss}^{k} + \ldots \tag{7.4}
$$

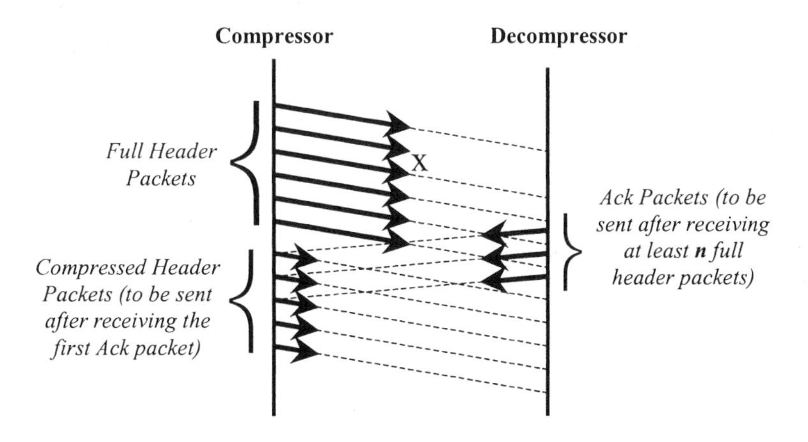

a) A successful case of HC context establishment

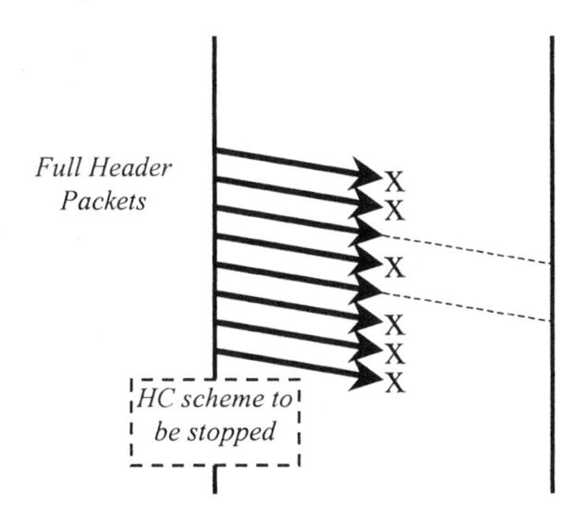

b) A failure case of HC context establishment

Figure 7-4. HC context establishment in the Ack mode with **m** = 4, **n** =3, **RTT** x $\mathbf{r_{pk}}$ = 4.

As the optimal mode operates at the condition of very low probability of packet loss, we expect that the compressor should be notified about the establishment failure after one or two attempts at sending the notice. If the probability of packet loss is higher, the HC context should be built according to the acknowledged mode as follows. After sending $\boldsymbol{m}$ full header packets, the compressor will start sending compressed packets only if it receives *Acknowledgement (Ack)* from the decompressor. Taking into account the $\boldsymbol{RTT}$ between the compressor and decompressor, the number of full header packets to be sent is now $\mathbf{Z} = (\boldsymbol{RTT} \times \boldsymbol{r_{pk}} + \boldsymbol{m})$ instead of $\boldsymbol{m}$. Fig. 7-4 shows an example of the acknowledged mode with $\boldsymbol{m} = 4$, $\boldsymbol{n} = 3$, and $\boldsymbol{RTT} \times \boldsymbol{r_{pk}} = 4$. The establishment cost can be calculated as follows

$$C_{ack} = Z(f - c) + n_{ack} s_{ack} \tag{7.5}$$

where $\boldsymbol{s_{ack}}$ is *Ack* message size, and $\boldsymbol{n_{ack}}$ is number of sent *Ack*s. As the decompressor starts to send *Ack* after receiving $\boldsymbol{n}$ or more full header packets, we can define $\boldsymbol{n_{ack}}$ as follow

$$n_{ack} = \frac{\sum_{k=1}^{Z} k p_{loss}^{Z-k} (1 - p_{loss})^k}{\sum_{k=1}^{Z} p_{loss}^{Z-k} (1 - p_{loss})^k} - (n-1) \tag{7.6}$$

The numerical example in Fig. 7-5 presents establishment cost for the following scenario of voice over IP over a typical cellular link: round trip time between the compressor and decompressor $\boldsymbol{RTT} = 120$ ms; packet rate $\boldsymbol{r_{pk}} = 50$ pkts/sec as typical packet interval for compressed voice is 20 ms; full header size $\boldsymbol{f} = 84$ bytes if IPv6/UDP/RTP protocol stacks are used to carry voice packets; compressed header size $\boldsymbol{c} = 1$ byte as we just need to carry one byte CID in it; a typical voice payload $\boldsymbol{d} = 30$ bytes; notice and *Ack* message size $\boldsymbol{s_{notice}} = \boldsymbol{s_{ack}} = 20$ bytes.

As can be seen from the graphs, the establishment cost in the optimal mode is significantly less than the establishment cost in the Ack mode, because in the optimal mode the HC scheme can transit to the compressed state without Ack. At first, one may be unimpressed by the saving of a few hundred bytes achieved thanks to a transfer of the HC context from the old AR to the new AR. However, we should notice that this is a saving in the amount of data transmitted during the handover period, when the MN and AR require significant bandwidth for signaling related to registration process. Moreover, the saving becomes more significant in WLAN "hot spots" where large numbers of MNs may be roaming and require frequent handovers.

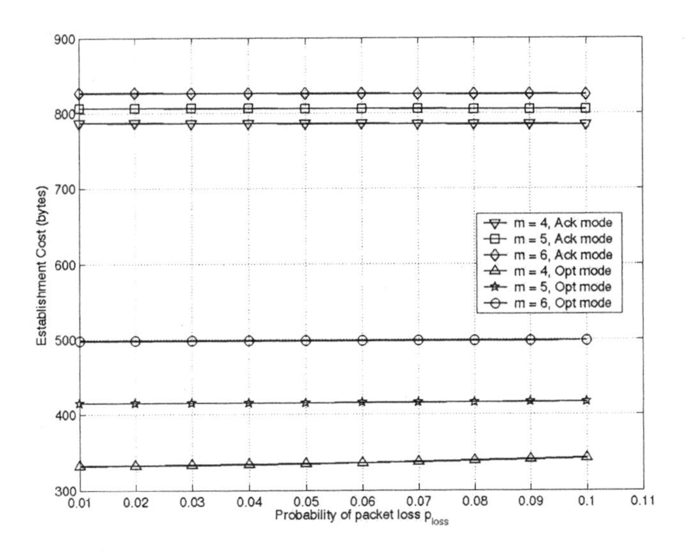

Figure 7-5. Establishment cost in the optimal (Opt) and acknowledged (Ack) modes. The required number of full header packet for context establishment ***n*** is 3.

2.3 Overview of Mobile IP

Mobility in the Internet has recently been a very active research topic within the IETF community. Intense research effort has been devoted to the issue, and led to a development of protocol mechanisms in support of mobility. The IETF Mobile IP Working Group has developed a solution officially named **IP mobility support** but popularly called **Mobile IP** [2]. The main characteristics of Mobile IP include transparency to applications and transport layer protocols, and scalability. The main functions of Mobile IP include mobile host registration and tunneling of datagrams to mobile hosts. For the purpose of this chapter, we will briefly describe the registration procedure illustrated in Fig. 7-6. Whenever an MN discovers that it has moved into a new subnet, it sends a *Registration Request* message, which includes the MN home address, care of address and Home Agent (HA) address, to the new Foreign Agent (FA). The new FA, located at the new Access Router, relays the message to the HA after retrieving information necessary for serving the MN in the future.

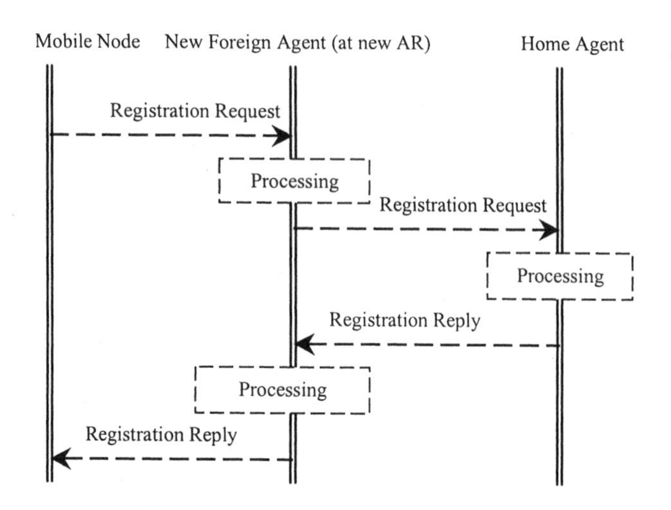

Figure 7-6. Mobile IP registration process

In response to the *Registration Request,* the HA sends a *Registration Reply* to the new FA. In turn, the FA sends the *Registration Reply* to the MN to confirm (or reject) the MN registration at the new FA. Following this registration procedure, all communications to and from the MN are routed via the new FA. For full description of Mobile IP protocol, readers should refer to [2].

2.4 Problem Description

Let us consider a scenario where the MN is running a Voice over IP (VoIP) application session, and the HC context was established at both MN and AR to allow transmission of compressed voice packets. When the MN performs a Mobile IP handover to the new FA, the HC context must be re-established, as the new FA has no knowledge of the previously transmitted packets i.e. the HC context. The context reestablishment, which involves sending ***m*** packets with full headers, is a time- and bandwidth-consuming process as we can see from the subsection 2.2. In another example, the cost of the HC context establishment in a shared channel such as Wireless LAN can be 26.4% extra channel usage by the full packets, and therefore reduction in the effective compression ratio by more than 7% [3]. Therefore, a potential enhancement is to transfer the HC context from the old AR to the new AR to avoid context re-establishment. However, this approach fails if packets are lost during handover. As the HC context depends on previous packets received, any packet loss can lead to miss-synchronization between compressor and decompressor. A miss-synchronization will force the compressor and decompressor to establish a new HC context, from the beginning (i.e. from the Initial State in Fig. 7-2). Unfortunately, packet loss often does take place in the handover process. Even though the buffering techniques can reduce packet loss, there is no guarantee that packet loss can be avoided altogether. Moreover, buffering may increase packet delay, which affects time-sensitive application such as VoIP [11]. Therefore, in Mobile IP networks it is important to perform HC context transfer that avoids the miss-synchronization problem, which may negate any benefits of context transfer.

3. PROPOSED SOLUTION

In this section, we propose a method for avoiding the miss-synchronization problem in HC context transfer (Section 3.1), as well as a method for efficient transfer of context from the old AR to the new AR (Section 3.2).

3.1 Avoiding miss-synchronization problem

Our solution is derived from a simple observation of the three-state HC scheme, described in Section 2.1. Since the HC context in the 2nd order state is very sensitive to packet loss (i.e. a lost packet results in a return to the Initial State), the compressor and decompressor should return to the 1st order state whenever there is an indication of an upcoming Mobile IP handover (for example, link layer triggers can be used as an indication). Readers will recall that in the 1st order state the HC context is made up of only STATIC fields; therefore, packet loss during handover time will not affect the HC context. The HC context of the 1st order state is to be transferred by the method specified in Section 3.2. When the handover process is completed, the HC may return to the 2nd order state. The avoidance of miss-synchronization is achieved at the expense of lower compression ratio in the 1st order state. However, in the following, we will show that this is a reasonable trade-off, since there are only small differences between the compression ratios in the 1st order state and the 2nd order state.

Table 7-1 shows the contribution to overhead reduction of the 1st and 2nd order states for different packet headers. The following notations are used

s: The total size of STATIC fields.

d: The total size of DELTA fields.

Δ: The total size of fields carrying small values (delta). Usually, one byte field is enough to carry delta value of a DELTA field.

$\boldsymbol{R_1}$: Overhead reduction of the 1st order state compared with the initial state.

$\boldsymbol{R_2}$: Overhead reduction of the 2nd order state compared with the 1st order state.

All values are expressed in bytes.

We can see from Table 7-1, that for all protocols (except TCP) the compression ratio in the 1st order state is only slightly less than in the 2nd order state. For example, when using IPv4, the header is reduced by 14 bytes in the 1st order state, and only by a further 1 byte when moving from the 1st order state to the 2nd order state. When using TCP, although the 1st order state brings a reasonable overhead reduction, operating in the 2nd order state significantly enhances the compression ratio. However, for VoIP applications TCP is not the preferred transport protocol.

If we assume that the CID requires 1 byte in the IP header field of a compressed packet, then Table 1-2 shows the overhead reduction for a VoIP application that uses RTP over UDP over IP (v4 and v6). $\boldsymbol{R_1}$ indicates the reduction when operating in the 1st order state, while $\boldsymbol{R_{overall}}$ indicates the reduction when in the 2nd order state. Intuitively, we can see that the small reduction in compression ratio when reverting to the 1st order state during the handover process is a good trade-off for avoiding miss-synchronization (and hence full context re-establishment).

Table 7-1. Contribution (in bytes) to overhead reduction

Packet Header	s	d	Δ	Contribution to $\boldsymbol{R_1}$ (s)	Contribution to $\boldsymbol{R_2}$ $(d-\Delta)$
IPv4	14	2	1	14	1
IPv6	38	0	0	38	0
TCP	4.5*	12	2	4*	10
UDP	4	0	0	4	0
RTP	4.5*	6	2	4*	4

* The compressed header size is an integer number of bytes

Table 7-2. Examples of overhead reduction (bytes) in three-state Header Compression scheme

Packet Header	Contribution to $\boldsymbol{R_1}$	Contribution to $\boldsymbol{R_{overall}}$
IPv4/UDP/RTP	21	26
IPv6/UDP/RTP	45	49

3.2 Piggybacking method of Header Compression Context Transfer

Assuming that miss-synchronization can be avoided, for the HC context transfer to be beneficial we require means for transferring the context from the old AR to the new AR with minimum overhead. We propose that the MN informs the new AR of the HC context during the handover procedure. This is based on the observation that the HC context of the compressor (or decompressor) at the old AR is the same as the HC context of the decompressor (or compressor) at the MN. In other words, the MN has a copy of the HC context held at the old AR; when the handover process begins, the MN will transfer the HC context to the new AR. For instance, the MN can append the context data to the *Registration Request* message in the Mobile IP protocol as a *Registration Extension* (which we will call *Context Transfer Extension*). At the new AR, the *Context Transfer Extension* will be detached from the *Registration Request* message and the contents of the *Context Transfer Extension* can be stored as the HC context of the link between the new AR and the MN. The rest of the *Registration Request* message will be processed according to the normal Mobile IP procedures. This method does not require any new signaling messages between network entities such as the MN, old AR and new AR. It also guarantees that the transferred HC context is available at the new AR before the MN and the new AR begin to exchange packets within the HC session. The *Context Transfer Extension* containing context data of the 1st order state will be of a few tens of bytes in size (for example, the context size for IPv4/UDP/RTP is 23 bytes). With the typical *Registration Request* message size of a few hundred bytes, the *Context Transfer Extension* is just a small addition.

4. SIMULATION RESULTS

We have used the OPNET simulation tool (http://www.mil3.com) to validate the proposed solution and to compare its performance with the case of full HC context re-establishment at the new AR.

Fig. 7-7 illustrates the network topology that has been used in the simulation. The border routers R_F, R_H and R_C connect the foreign domain, home subnet and corresponding subnet to the Internet. The HA and FAs reside in ARs equipped with IEEE 802.11 access point functionality. The Mobile IP functionality has been implemented in the HA, FAs and MN using the simulation model developed by T. Park [12]. The acknowledgement-

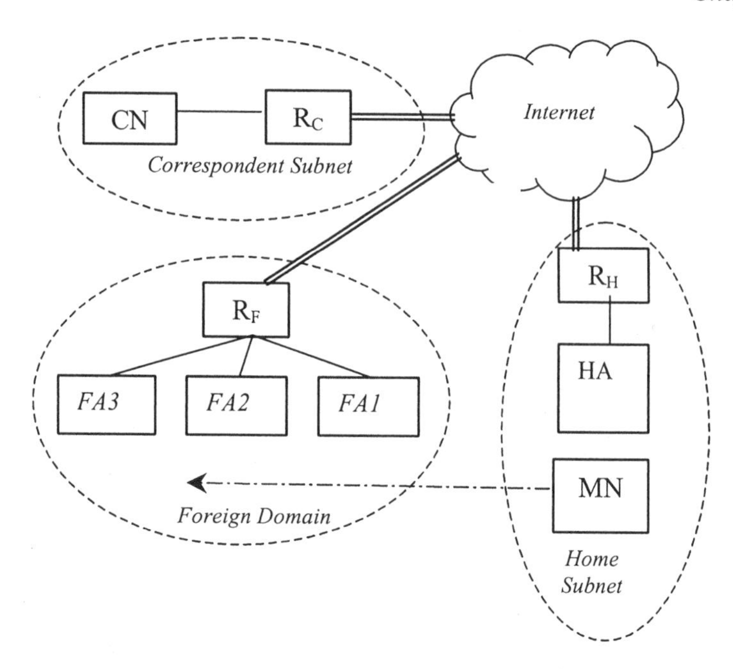

Figure 7-7. Network topology used in the OPNET simulation environment

based three-state HC scheme was implemented at the HA, FAs and MN. The HC context establishment requires the compressor to send ***m*** full header packets (for the results presented in this chapter we assumed ***m*** = 3) and to receive an Ack from the decompressor. For simplicity, we only model the IPv4/UDP protocol stack to carry voice packets. To obtain the constant data rate, we disabled silence compression of the voice encoder (i.e. voice packets are still generated during the silence period). The constant data rate makes simulation observations and analysis easier. VoIP application uses G.729 as the encoder scheme, with frame size equivalent to 10 ms and coding rate of 8kbps.

The following scenario has been used in the simulation: the MN connects with its HA at time $t = 0$ s, and starts a VoIP session with the Corresponding Node (CN) at the time $t = 15$ s. While the VoIP session is ongoing, the MN begins to move away from its home subnet toward the foreign domain in a horizontal linear path with constant speed of 30 kms/h. As the MN moves across the foreign domain, it will encounter handover with FA1, FA2 and then FA3 at the times, approximately, t = 26.1 s, 86.1 s and 146 s. The packet loss during handovers causes a context miss-match between the compressor and decompressor, forcing context re-establishment.

We also investigated how our HC context transfer protocol performs during the handover periods. It is worth mentioning that since VoIP is a two-way interactive application, packets of the flow (CN→ MN) are compressed at FA or HA and decompressed at MN; meanwhile MN acts as compressor and FA (or HA) acts as decompressor for packets of the flow (MN→ CN). Fig. 7-8 shows a "snapshot" of packet size versus time at the MN compressor during handover periods with FA1, FA2 and FA3. In the case of the handover without context transfer, after sending a Mobile IP *Registration Request* (Point 1 in Fig. 7-8), the MN has to send full header packets (Point 2 in Fig. 7-8) to establish a new HC context for the flow (MN→ CN). An *Acknowledgment* (Point 3 in Fig. 7-8) of context establishment for the flow (CN→ MN) is also sent. On the other hand, in the case of handover with context transfer, the MN sends Mobile IP Registration Request with *Context Transfer Extension* (Point 1 in Fig. 7-8), and then the MN can send compressed header packets to the new FA. MN also does not need to send *Acknowledgment* for the flow (CN→MN) as the new FA, with HC transferred context, can start sending compressed header packets immediately.

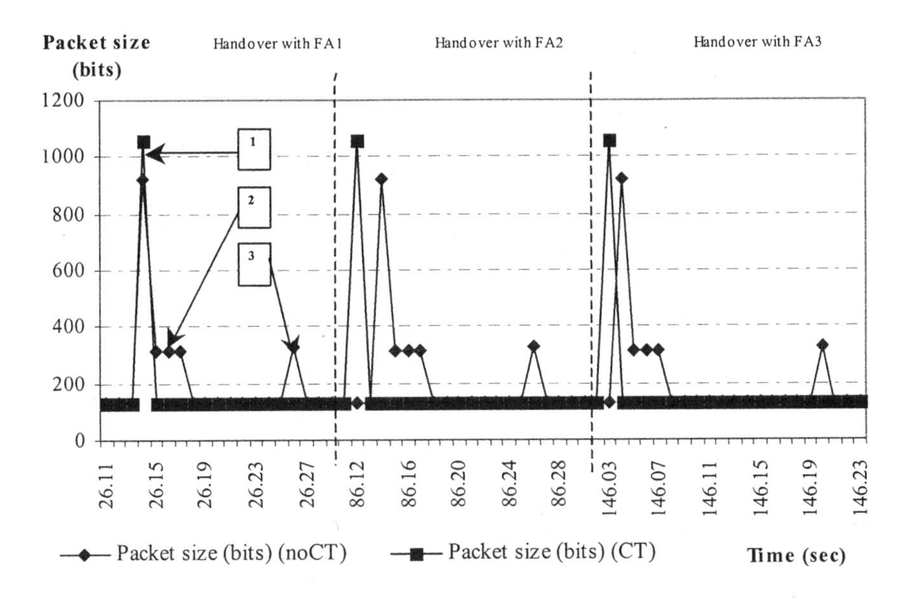

Figure 7-8. Packet size at MN compressor in two cases: with Context Transfer (CT) and without CT

The same behavior is observed for the handovers with FA2 and FA3 (the times when handovers occur differ slightly between the graphs because each scenario is run as independent simulation). In all handover periods, based on the above packet size behavior, we can observe that the MN without context transfer generates more traffic load than the MN with context transfer. As discussed in 2.2, the reduction in load brought in by the context transfer depends on factors such as the number of full header packets sent to establish the HC context, whether or not the Acknowledgements are used, the ratio of packet payload to packet header, and the round trip time (***RTT***) between compressor and decompressor. For the scenario assumed in our simulations, we estimate that approximately a 20% load reduction on the WLAN link is achieved during the handover period. This reduction is quite significant if we consider that the MN and AR usually require more bandwidth for Mobile IP signaling during the handover. Although the handover period may be relatively short (e.g. a few hundred milliseconds) the load reduction becomes a significant factor in "hot spots" where WLANs are serving large numbers of MNs encountering frequent handovers.

5. CONCLUSION & FUTURE WORK

In this chapter, we presented a simple solution to the problem of miss-synchronization, which occurs due to packet losses during handover when transferring header compression context in Mobile IP networks. In addition, we have illustrated how to transfer the HC context from an old AR to a new AR using existing Mobile IP signaling messages. Preliminary simulation results indicated that load reduction of 20% could be achieved during the handover period by transferring the HC context. The current work can be extended by applying our approach to a cellular mobile network link. It is expected that the HC context transfer in a cellular network can bring even higher benefits, due to the higher cost of HC establishment in the cellular link. The main reason for this is that as the cellular link has much longer ***RTT***, typically 100-200ms, there are more in-flight full header packets during HC establishment period. Also, during handover period more in-flight compressed packets may need to be discarded at the decompressor if miss-synchronization between compressor and decompressor occurs. This makes using the proposed HC context transfer more beneficial.

REFERENCES

1. C. Bormann (editor), "Robust Header Compression (ROHC)," Request for Comments 3095, IETF, July 2001.
2. C. Perkins (editor), "IP Mobility Support for IP v4," Request for Comments 3320, IETF, January 2002.

3. C. Westphal and R. Koodli, "IP Header Compression: A Study of Context Establishment," in *Proceedings of IEEE Wireless Communications and Networking Conference (WCNC2003)*, New Orleans, LA USA, 2003, vol.2, pp. 1025-1031.
4. Context Transfer, Handoff Candidate Discovery, and Dormant Mode Host Alerting (Seamoby) Working Group, IETF, Http:// www.ietf.org/html.charters/seamoby-charter.html [On-line access 4 September 2003]
5. J. Kempf et. al., "Problem Description: Reasons For Performing Context Transfers Between Nodes in an IP Access Network, " Internet Draft of IETF, draft-ietf-seamoby-context-transfer-problem-stat-04.txt (work in progress), December 2001.
6. J. Loughney et. al., "Context Transfer Protocol," Internet Draft of IETF, draft-ietf-seamoby-ctp-00.txt (work in progress), October 2002.
7. M. Degermark, B. Nordgen, and S. Pink, "IP Header Compression," Request for Comments 2507, IETF, February 1999.
8. R. Koodli R. and C. E. Perkins, "A Context Transfer Protocol for Seamless Mobility," Internet Draft of IETF, draft-koodlt-seamoby-ct-04.txt (work in progress), August 2002.
9. R. Koodli R. and C. E. Perkins, "Fast Handovers and Context Transfers in Mobile Networks," in *ACM Computer Communication Review*, vol. 31, number 5, 2001.
10. S. Casner and V. Jacobson, "Compressing IP/UDP/TCP Headers for Low-Speed Serial Links," Request for Comments 2508, IETF, February 1999.
11. T. Park and A. Dadej, "Adaptive Handover Control in IP-based Mobility Networks," in *Proceedings of 1st Workshop on the Internet, Telecommunications and Signal Processing (WITSP 2002)*, Wollongong, Australia, 2002,pp 34-39.
12. T. Park and A. Dadej, "OPNET Simulation Modeling and Analysis of Enhanced Mobile IP," in *Proceedings of IEEE Wireless Communications and Networking Conference (WCNC2003)*, New Orleans, LA USA, 2003, vol.2, pp. 1017-1024.
13. V. Jacobson, "Compressing TCP/IP Headers for Low-Speed Serial Links," Request for Comments 1144, IETF, February 1999.

Chapter 8

HANDOVER CHANNEL ALLOCATION BASED ON MOBILITY PREDICTIONS

Aruna Jayasuriya
Institute for Telecommunications Research, University of South Australia, Mawson Lakes SA 5095, Australia

Abstract: Future cellular communication networks must use superior handover processing techniques to guarantee that continuity of connections during handover. Although previous studies have shown that the blocking handover probabilities can be reduced by exclusively reserving resources for handover users, this mechanism only results in marginal increase in overall system performance. This is due to the increased blocking probabilities suffered by new users as a result of channel being permanently allocated for handover users, even when no handover user requires them. In this paper we propose a method to dynamically allocate the handover channel after estimating the number of handover channels required by impending handover users at any given time. We use a mobility model to predict the user's handover probabilities and a dynamic resource allocation algorithm to estimate the number of channels required to support the predicted handover traffic. Results of the study show that the proposed solution increases the system utilisation up to 5%.

Key words: mobility predictions, advance resource allocations, handover

1. INTRODUCTION

Future mobile communication networks promise to offer services such as video conferencing and e-commerce applications, which require a higher guarantee from the networks that the services can be continued uninterrupted during handover; i.e. seamless handover. Lower blocking probabilities can be achieved by reserving a set of channels exclusively for handover users

[1,2]. However it has been shown that only marginal gains in system performance can be achieved through this scheme [1,2]. Furthermore it was also observed that with this scheme new users suffer very high blocking probabilities even at lighter loads leading to degradation in overall system performance [1,2]. Therefore this reservation scheme does not seem to be an effective solution to provide lower blocking probabilities for handover users in future mobile communication networks. Goal of this study is to achieve lower blocking probabilities for handover and new users at higher system utilisations. We propose an innovative scheme, using user mobility information to optimise the resource reservation in cellular networks. The proposed handover strategy consists of two major components:

1. Mobility Prediction Model - A scheme that predicts the most probable target cells and the probabilities of handing over to these cells, dependent on the user's current state.
2. Advance Allocation Algorithm - An allocation strategy that determines the number of handover channels required in any cell, given the current state of the system and current position of the mobile users.

2. MOBILITY MODELS

In this study, mobility models were used to estimate the amount of handover traffic expected at any cell at any given time. The mobility model used is based on the mobility patterns observed in the network coverage area [3]. A user's position, speed and direction of movement, (collectively called a *mobility vector* throughout this paper) over a time period were used to describe a user's behaviour. The mobility model used in this work can estimate the probability of a user making a handover to any of the neighbouring cell given the user's current position, speed and direction of movement. Due to space limitation we have not included a detailed description of the mobility model in this paper. Detailed information on the mobility prediction model used in this work is presented in chapter 6 of [2] and in [3].

3. SERVICE CATEGORIES

Future mobile communication networks intend to provide a variety of services with different Quality of Service (QoS) levels [4,5,6]. Video conferencing, e-commerce, high quality data service, remote sensing, video on demand and tele-medicine are just a few of the new services promised by third generation networks. All these different services are also expected to receive different Grade of Service (GoS) levels. GoS parameters studied in this investigation were the blocking probabilities for new and handover users and handover delay. In this work different services are classified under different priority levels. These priority levels determine the GoS a particular service can expect from the network. We have selected three service classes namely, *video*, *voice* and *data* with the highest priority being given to the video service class and lowest to the data class.

All the services are assumed to be Constant Bit Rate (CBR) type services with the data service class being able to tolerate considerable handover delay or call set up delay. In this paper call set up delay refers to any delay experienced by new users trying to gain access to the network. The other two service classes are assumed to be unable to tolerate any handover or call set up delay. Readers should not confuse these service classes with the actual video, voice and data services described in common QoS documentations. These names have been used for the service classes to convey the concept that these types of services will be available in the third generation networks.

We have selected the transmission parameters as listed in Table 8-1. These values allow us to demonstrate the provision of various future services with different QoS levels. The rate X is the Bandwidth Unit (BU) or the effective rate when the user transmits data in one slot per frame.

Table 3-1. Transmission parameters for the three services.

Service Type	Priority	Transmission rate
video	1	$3X$
voice	2	X
Data	3	$2X$

4. NETWORK MODEL

In this section we describe the network model used to demonstrate the performance improvement of the handover process, achieved by employing a handover strategy based on the mobility predictions. Fig. 8-1 shows the network model used in this study. Mobile Terminals (MTs) and Cell Site Switches (CSSs) in this network contain functional entities to record mobility patterns and make mobility predictions based on these patterns. This adds a considerable number of new procedures to the system, requiring substantial computational capacity. Hence the simple network model given in Fig. 8-1. was selected to reduce the computational resources and time required for modelling and simulating of the network. Users in the network can be either stationary or follow one of the mobility patterns shown in Fig. 8-1.

MTs periodically transmit the received signal strength measurements with respect to different Base Transceiver System (BTS) to the BTS they are currently connected to. The BTS is able to convert the signal strength measurements into spatial data which is sent to the CSS for further processing. Signal strength measurements are averaged before this conversion to smooth out any sudden changes in the data. Procedures used to convert signal strength measurements into spatial information are described in [3].

When the network is in operation it starts gathering mobility information and building the mobility model. Once enough mobility data has been gathered, this model can be used to make handover predictions as described in [3]. During the course of operation more mobility data are gathered and the existing mobility model is updated periodically to introduce the effect of new data. This is important as changes in vehicular traffic patterns are quite common in urban environments. However if the mobility patterns of the users do not change over a long period, the subsequent updates do not add to the accuracy of the model.

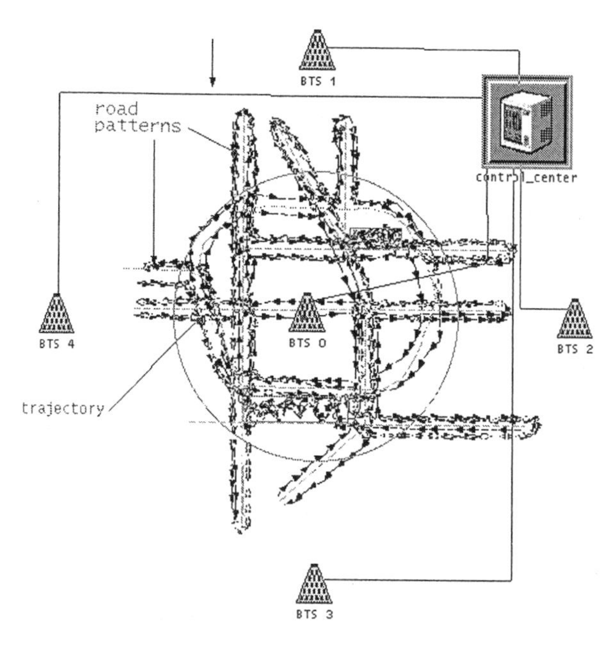

Figure 8-1. Mobile Network for collecting mobility data.

The majority of the mobility predictions and handover processing is carried at the Cell Site Switch (CSS), shown as control_centre in Fig. 8-1. The cell site switch enables the centralised control of a large number of base stations under its control. It also maintains databases for the base stations and makes handover decisions on behalf of the base stations.

5. ADVANCE ALLOCATION ALGORITHM

It has been shown that channels exclusively reserved for handover users are not required most of the time of network operation [1]. They only become necessary when a handover user tries to enter a congested cell. The aim of this work is to estimate the number of channels required for imminent handover users at a given time. This is addressed by the *Advance Allocation Algorithm*, where some resources are reserved in advance for highly loaded cells in anticipation of imminent handover by users. The number of resources reserved in the target cells are determined by the aggregate of estimated probabilities of all users in neighbouring cells making a handover to the particular target cell. Rather than pinpointing the target cell, the algorithm selects a set of probable target cells and estimates the probability of making a handover to these cells. These probabilities are estimated using the mobility models presented in [3].

Advance Allocation Algorithm is used to reserve resources in advance for users that are predicted to make handovers to probable target cells. The flow diagram given in Fig. 8-2 shows the *Advance Allocation Algorithm* proposed in this research (r in Fig. 8-2 refers to the distance between mobile terminal and current base station). This algorithm will be explained in detail in the remaining part of this section. The thick dashed lines between different databases and the flow chart processes indicate an information flow between databases and the algorithm processor. The direction of the information flow is indicated by the arrowhead.

After extracting the mobility information from signal strength measurements, BTS passes the mobility information as well as the signal strength measurements to the CSS. Although the handover predictions are carried out based on the mobility information, it has been observed that the precise instance for the handover is better determined by the signal strength measurement [7].

Due to the local variations of the signal strength measurements caused by shadowing and fading effects it is very hard to define a boundary for a cell where the handover has to be carried out. It can also be time dependent for a cell due to the movements of different obstacles in the area. Therefore it is safer to execute the handover based on the signal strength using the following rule:

If $S_{\text{current}} < S_{\text{HO}}$ and $(S_{\text{cell}(i)} > S_{\text{current}})$
{handover to cell i}

where S_{current} is the signal strength measurement from the current cell, $S_{\text{cell}(i)}$ is the signal strength measurement from cell i and S_{HO} is the signal strength required for maintaining a connection with the required quality.

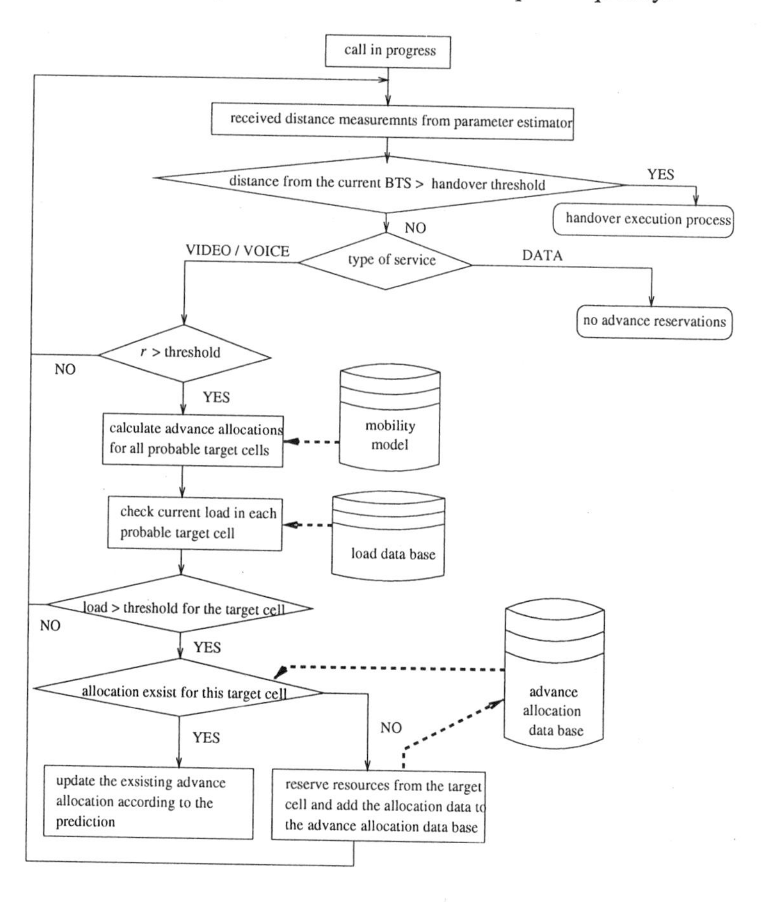

Figure 8-2. Advance Allocation Algorithm.

This instructs the system to perform a handover if the current signal strength is below the signal strength required to maintain a reliable connection and the signal strength from the current base station is not the maximum of the signal strengths received from all base stations in the vicinity of the mobile node. For voice and video users, if a handover is required, the handover execution process will be carried out as highest priority. If there are not enough resources to support the voice and video handover users, they will be blocked.

In section 3 we mentioned that the data users are able to tolerate some delay. Therefore, the advance allocation strategy was not used for data users. Instead data users will be inserted into a queue where they can wait up to 45 seconds while enough resources are found to support them. 45 seconds has been identified as the maximum time users are prepared to wait before terminating the connection [5]. As we discussed earlier, handover users should always get priority over new users in the system. Therefore in the data queue, requests from handover data users are inserted before the new data users. Fig. 8-3 shows the policy we used to queue the data users.

Once any user terminates a call or makes a handover to a neighbouring cell, the *handover manager* checks whether there is enough resources in the cell to support a data user. If so it takes the request at the head of the data queue and takes the necessary steps to allocate enough resources for the user. A timer for 45 seconds is started with every data request inserted into the queue. The timer for a particular user is cancelled once enough resources have been allocated. If not, once the timer expires the request is removed from the queue and this results in handover failure (for handover data users) or new call failure (for new users).

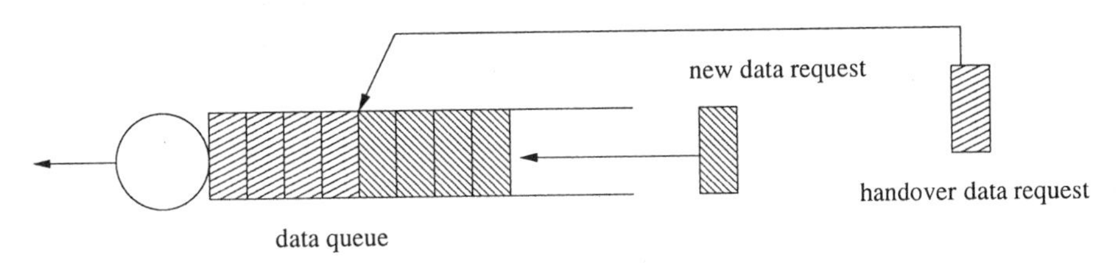

Figure 8-3. Queue for new and handover data users.

If handover is not required, after examining the received signal strength measurements, the algorithm determines whether it is necessary to perform any advance allocations for the users. If not, as in case of data services, packets are sent to the correct destination.

5.1 Effective Allocation Zones

For voice and video connections, based on the current position of the mobile terminal and the current state of the system, the *advance allocation algorithm* decides whether allocations of handover only channels are necessary for a particular user. Unnecessary allocation of handover only channels reduces the operating capacity of the system by creating non-utilised channels. Further it is undesirable to request an allocation too early before the mobile makes a handover to a target cell. The solution to this is to wait until the mobile gets relatively close to the boundary of the cell before making allocations. Therefore we have decided to make advance allocations only when the mobile terminals are at a distance greater than r_{alloc} from the current base station.

The easiest way to define the cell boundary is to define a circle with distance r_{cell} from the base station, leading to a circular handover allocation boundary with radius $r_{\text{alloc}} = kr_{\text{cell}}$. Where k is a value between 0 and 1. However it has shown that is very hard to define the boundary of a terrestrial cell in terms of the distance form the base station, due to shadowing effects [7]. Therefore, due to heavy shadowing effects, it is possible that the actual cell boundary in a certain area is smaller than the r_{alloc} selected for the cell. This leads to users in those areas missing out on the advance allocation procedures, resulting in a possible degradation of overall performance. To resolve this problem we have decided to use a relatively small value for k, thereby reducing the probability of excluding any area from the region subjected to advance allocations. In [3] it was argued that it is not necessary to predict the behaviour of users when they are close to the centre of the cell and selected $r > 0.65r_{\text{cell}}$ as the effective region for a mobility vector to be included in the mobility model. Therefore it was decided to use the advance allocation on users who are more than $0.65r_{\text{cell}}$ from the current base station.

5.2 Allocation Function

The goal of the advance allocation strategy is to improve the probability of handover users finding enough resources in the target cells to continue their service. During the allocation phase there is not enough information to accurately predict the target cell for the mobile terminal. Instead the probability of the user making a handover to each of the neighbouring cells can be estimated. Based on this information it is not justified to set aside resources in the most probable target cell, to be used by a particular user. The user may terminate the call before making any handover or may handover to another cell leading to a wasted allocation. To minimise the risk of making an allocation in a wrong target cell, allocations can be made in more than one cell. Although this increases the chances of a user finding resources in the target cell, it also leads to a large degradation of operating system capacity due to the large number of under-utilised resources in some cells. This can be overcome by introducing a mechanism, which reserves a certain fraction of the required resource into a pool of handover only channels. Then any handover user can obtain the required resources from this pool and continue service in the target cell. Allocation of the resources into a pool, increases the multiplexing gain of the system by allowing all the handover users to contend for a certain amount of resources.

The amount of resources added to the pool of handover only channels is determined from the estimated probability of the mobile user making a handover into the cell. This value is readily available through the mobility model [3]. In [3] it was explained that mobility vectors are rounded to reduce the amount of storage required to store the mobility models. The rounding is done according to the following rules:

- angular position - to the nearest 5 degrees.
- direction - to the nearest 5 degrees.
- speed - to the nearest 4 ms^{-1}.

This rounding also reflects the fact that these values cannot be accurately measured in a real network environment due to non-linear fading effects. Mobility vectors used for handover predictions are also rounded according to the above rules. The probability of handing over to a particular cell can then be read directly from the mobility model for that cell. Assume, that the probability of handing over to cell n is $P_n(\mathbf{M})$, where $\mathbf{M}$ is the mobility vector for the current state of the mobile user. Then the advance allocation for cell n, ΔH_n, is determined from equation (8.1). This is referred to as the *allocation function* in the rest of the document. Allocation function is shown in Fig. 8-4.

$$\Delta H_n = \begin{cases} 0 & \text{if } P_n(\mathbf{M}) < 0.5 \\ (1.4P_n(\mathbf{M}) - 0.4)C & \text{otherwise} \end{cases} \tag{8.1}$$

where C is the number of channels required to support this connection.

The reasons for selecting this function will be explained later after describing the other key features of the algorithm. A new allocation is added to the total allocation that has been already made to this cell, HA_n. This total is then rounded up to find the number of channels that need to be set-aside for handover users. The number of handover channels allocated within cell n is denoted H_n. If there are a large number of users predicted to make handovers to a particular cell, a large portion of this cell's resources may be tied up through the allocation procedure. This creates a highly undesirable environment for new users and results in degradation of the system operating capacity. Therefore the maximum number of resources that can be set aside for handover users has to be limited to a suitable number of channels, $H_{\max}$. The allocations are added to the pool only if $HA_n < H_{\max}$. After addition HA_n is modified as follows:

$$HA_n = \begin{cases} H_{\max} & \text{if } HA_n > H_{\max} \\ \mathrm{HA}_n & \text{otherwise} \end{cases}$$

Assume that the total number of channels available per cell is S and the current load of the cell n is L_n. Then if the load on cell n is more than $S - HA_n$ then the allocation amount is limited by the available channels. Therefore we set,

$$HA_n = \min(S - L_{n,} HA_n) \tag{8.2}$$

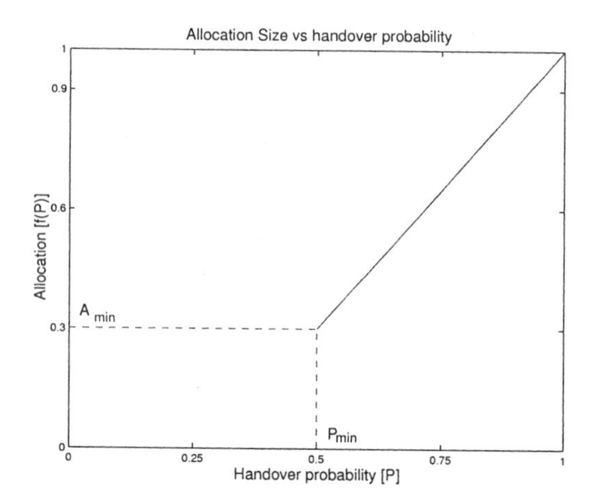

Figure8- 4. Allocation function.

Due to the rounding up of HA_n, the sum of very small allocations do not affect the number of total channels allocated for handover users. At the same time if the probability of entering a cell is very high this should be reflected in the allocations by reserving enough resources in the target cell. It was decided to allocate C channels for connections that have probability 1 of handing over to a particular cell. C is the number of channels required per frame to support that connection. This leads to selecting 1 as the upper limit of the function.

The other parameters used for the allocation function (equation (8.1)) were empirically decided. Due to the time required to run the final simulations we could not optimise all the parameters of the function. Therefore smaller simulation environments were set up to observe the impact of the allocation function on the total number of channels allocated for handover users. Following parameters of the allocation function were varied; lower limit or handover probability which warrants an allocation

($P_{\min}$) and the allocation amount at this probability ($A_{\min}$). The average number of handover channels allocated during each of these simulations was then noted. During these simulations we have selected $H_{\max}$ to be 4 channels allowing 3 channels for video users and 1 channel for voice users. To optimise the average number of handover channel allocations, $P_{\min} = 0.5$ and $A_{\min} = 0.3$ were selected which resulted in the allocation function given in equation (8.1).

It has to be emphasised here, that the simulations used for this estimation process only measured the average number of channels reserved by the algorithm. Performance of the algorithm, in terms of the blocking probabilities suffered by users, were not analysed at this stage. Therefore it cannot be stated that the above selected parameters are optimum. It should be noted that the function of the algorithm is to get an indication of the requirement of handover resources at a given time. It is not intended to find the exact number of channels required to support the handover users. Furthermore the selected parameters are suitable for the particular traffic patterns considered in the study. With different traffic patterns other values may result in better system performance. As the performance of the system is dependent on the average number of channels allocated and not the function shape, it is believed that fine-tuning of the parameters will have only limited impact on the final results.

If the target cells are lightly loaded the probability of finding enough resources for the handover users is much higher than that for heavily loaded cells. Subsequently, advance allocation strategies are not necessary to improve the performance of under-loaded cells. The need for advance allocation strategies arise when the load of the higher than a certain threshold. Therefore before allocations are made, the load of each cell is observed to check whether the current state of the cell warrants an advance allocation. In the subsequent simulations, 65% of the cell capacity has been selected as the load threshold which warrants an advance allocation.

The selection of these criteria do not have any bearing on the performance of the system as 65% of the system capacity is well below the normal operating region (Results in [1] suggest that the limiting operating utilisation is around 80). They simply reduce the amount of advance allocation processing.

5.3 Terminations and Handovers

Once a user terminates a call all the advance allocations included for this user have to be removed from the advance allocation database. Further the number of channels reserved for handover users in each of the target cell have to be updated to reflect the fact that these users no longer contribute to the handover only channel pool. Similarly after a handover the advance allocations for all the cells have to be removed and the handover only channel pool has to be updated. These procedures are shown in Fig. 8-5.

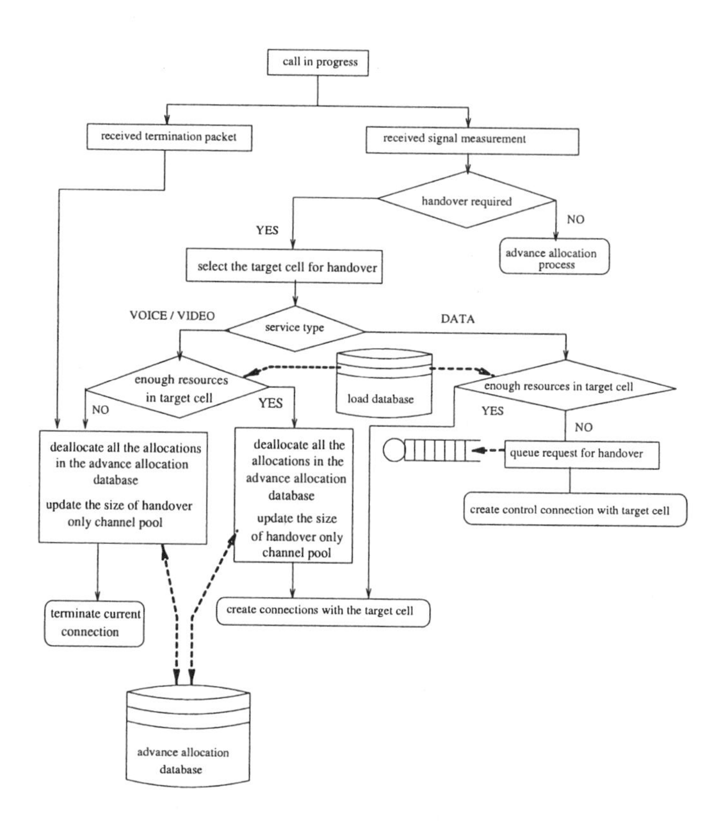

Figure 8-5. Deallocation procedure after terminations and handovers.

After terminations and handovers the data queue is searched to check whether there are any data users waiting in the queue. If there are enough resources left to accommodate these data users, channels will be allocated.

5.4 Blocking for New and Handover Users

The performance of the proposed allocation algorithm depends on the maximum number of channels that can be allocated under the advance allocation scheme. It should be recalled that a video user requires 3 channels and a voice user requires 1 channel. Therefore we have decided that to use 4 handover channels as the maximum that can be allocated for a handover user at a given time. The presence of three service categories and the use of handover only channels for some of these service types makes it harder to determine when a particular type of user is blocked by the network. With a maximum of 4 handover only channels, the following rules can be used to determine whether to admit a user of a particular type. Assume that the maximum number of channels available in a cell was 100. *L* is the current load of the cell and *HA* is the number of channels allocated for handover users.

- All data users, new video users and new voice users are blocked if $L \geq 100 - HA$.
- Voice handover users are blocked if $L = 97$ and L=100.
- Video handover users are blocked if $L \geq 98$.

We demonstrate in later sections through simulations that this is the *optimum* number of channels for the system.

5.5 Advance Allocation Database

MTs periodically send signal strengths measurements to the BTSs. When such a packet has been received at the CSS, it should know whether enough allocation has already been made for this connection. This is achieved by the use of an *Advance Allocation Database*. The advance allocation database consists of a list of allocations made by all connections in all probable target cells. After the first allocation for a connection, the algorithm creates a new record for this connection. In subsequent allocations the algorithm checks the advance allocation database to see whether there are any contributions from this connection towards the handover only channels for this cell. If so, the difference between the recorded allocations and the current allocations is added to the total allocation for that cell. This can be either positive or negative, leading to an increase or decrease in the handover channel pool. The new allocations then replace the old allocations in the advance allocation database.

6. PERFORMANCE OF THE PROPOSED ALLOCATION STRATEGY

In this section we present the results of the simulation studies. The blocking probability for new and handover users of voice, video and data services, system utilisation and queue delay for data users were used as the performance parameters in the simulation studies.

6.1 Simulation Scenarios

The network model given in Fig. 8-1 was used to evaluate the performance of the system. 1000 mobile users, equally divided between the three service types, were randomly scattered across the whole network.

The total number of channels available in each cell was limited to 100. The major objective of this simulation study is to show that the mobility prediction and the advance allocation algorithm improves the operating utilisation of the system. Performance of the proposed strategy was compared against a system with no handover only channel allocations and systems with a fixed number of channels allocated for handover users. Below is the list of simulations we used to evaluate and compare the performance of different strategies:

1. Simulate a network without allocating channels exclusively for handover users. Data users are allowed to wait in queue for 45 seconds.
2. Simulate a network with 4 channels allocated for voice and video handover users. Data users are allowed to wait in queue for 45 seconds. Priority given to handover data users over new data users in the data queue.
3. Simulate a network with 8 channels allocated for voice and video handover users. Data users allowed to wait in queue for 45 seconds. Priority given to handover data users over new data users in the data queue.
4. Simulate a network with 12 channels allocated for voice and video handover users. Data users are allowed to wait in queue for 45 seconds. Priority given to handover data users over new data users in the data queue.
5. Simulate a network with the prediction model and advance allocation algorithm which decides the number of handover only channels required at a given time. Maximum number of channels that can be allocated for handover users is 4. Data users are allowed to wait in queue for 45 seconds. Priority given to handover data users over new data users in the data queue.
6. Simulate a network with the prediction model and advance allocation algorithm which decides the number of handover only channels required at a given time. Maximum number of channels that can be allocated for handover users is 8. Data users are allowed to wait in queue for 45 seconds. Priority given to handover users data users over new data in the data queue.

The interarrival time was selected as an exponential distribution with mean of 5 minutes and the call length distribution was selected as an exponential with mean 3 minutes. The length of each simulation was 1200 seconds. For each simulation scenario 30 different simulation runs were carried out with 30 random seeds. This is equivalent to 10 hours of simulation time for each simulation scenario.

6.2 Blocking Probabilities with Different Allocation Schemes

In this section we present the blocking probabilities of various service categories under different allocation strategies. Fig. 8-6 shows the blocking probabilities for new and handover video users under various allocation schemes. Some terms in the legend in Fig. 8-6 have been abbreviated to improve the clarity of the figure. These abbreviations are given below.

1. new HO CH = n refers to new users, with n channels exclusively allocated for handover users.
2. handover HO CH = n refers to handover users, with system conditions as given in 1.
3. new predictions max HO CH = n refers to new users, with handover only channels are allocated by the advance allocation algorithm using mobility prediction. Maximum number of channels that can be allocated for handover users are 4.
4. new prediction HO CH = n new users with system conditions as given in 3.

Fig. 8-6 shows that blocking probability for handover video users can be reduced by allocating handover only channels. However the blocking probability for new video users increases rapidly when a large number of channels are allocated exclusively for handover users. For example at 73% system utilisation, (This is the system utilisation without any handover channel allocation.) the new user blocking probability increases by 8% with 8 handover only channels, compared to an increase of 2% with 4 handover only channels allocated. This rapid increase in new user blocking probability decreases the operating utilisation of the system. This was expected as the extensive analytical and simulation analysis of the basic network model showed that the optimum utilisation is always achieved with 1 channel allocated for handover users.

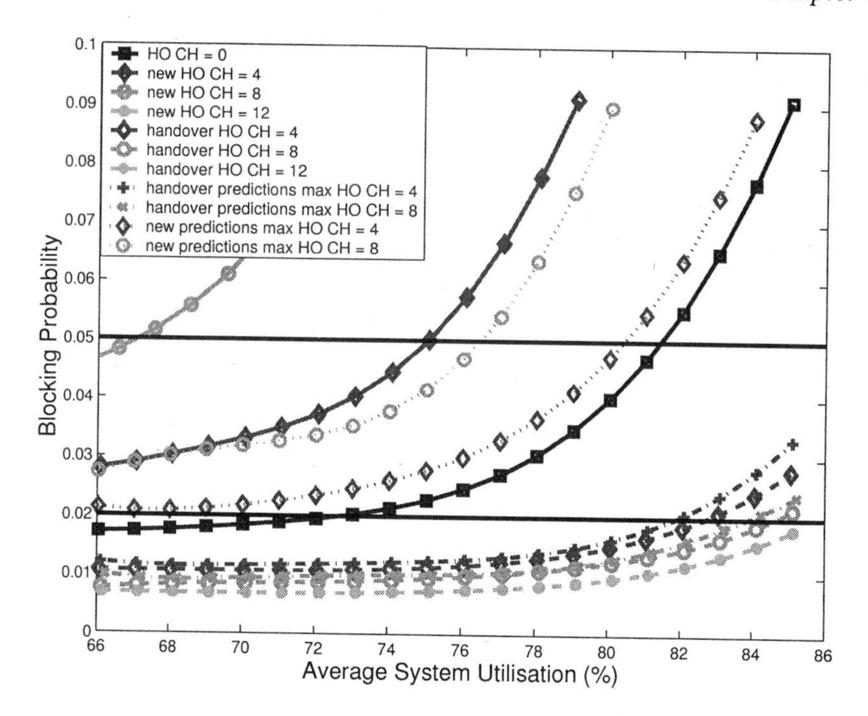

Figure 8-6. Blocking probabilities for video users with different allocation strategies.

The effectiveness of the advance allocation algorithm combined with the prediction model is evident from Fig. 8-6. It can be observed that, the new user blocking probability drops considerably with the proposed allocation scheme. Although the handover blocking probability has increased slightly, this increase is very minor compared to the decrease in the new user blocking probabilities.

It can be observed that, handover blocking probability increases only by 0.1% with the proposed scheme with maximum 4 handover only channels, compared with the permanent allocation of 4 handover only channels. However the new user blocking probability drops by 1.5% with the proposed scheme, compared with the permanent allocation scheme. The differences are more evident when the maximum number of handover channels was increased to 8. Blocking probability for new users drops by 6.8%, while the handover blocking probability increased only by 0.1%. This strongly suggests that, the proposed scheme provides the channels for handover users whenever they are required and new users have a far better chance of finding enough channels than for a system with permanent handover channel allocations. The results show that the proposed scheme minimises the number of channels unnecessarily allocated to handover users, as new users are far less disadvantaged than in the permanent allocation schemes.

Blocking probabilities for data users under various allocation strategies are shown in Fig. 8-7. Recall that data users cannot access the handover only channels. They may instead wait up to 45 seconds in a queue before being rejected by the network. Handover data users get priority over new data users in this queue.

Fig. 8-7 shows that the blocking probabilities for both new and handover users were considerably lower with the proposed allocation scheme, compared to the performance with permanent allocations. This drop in blocking probabilities was due to the higher availability of resources for data users, which results from the reduction of unnecessary handover channel allocations for voice and video users, by the proposed algorithm. This again confirms the effectiveness of the proposed method in reducing the blocking probabilities for all users that do not have channels allocated exclusively for them.

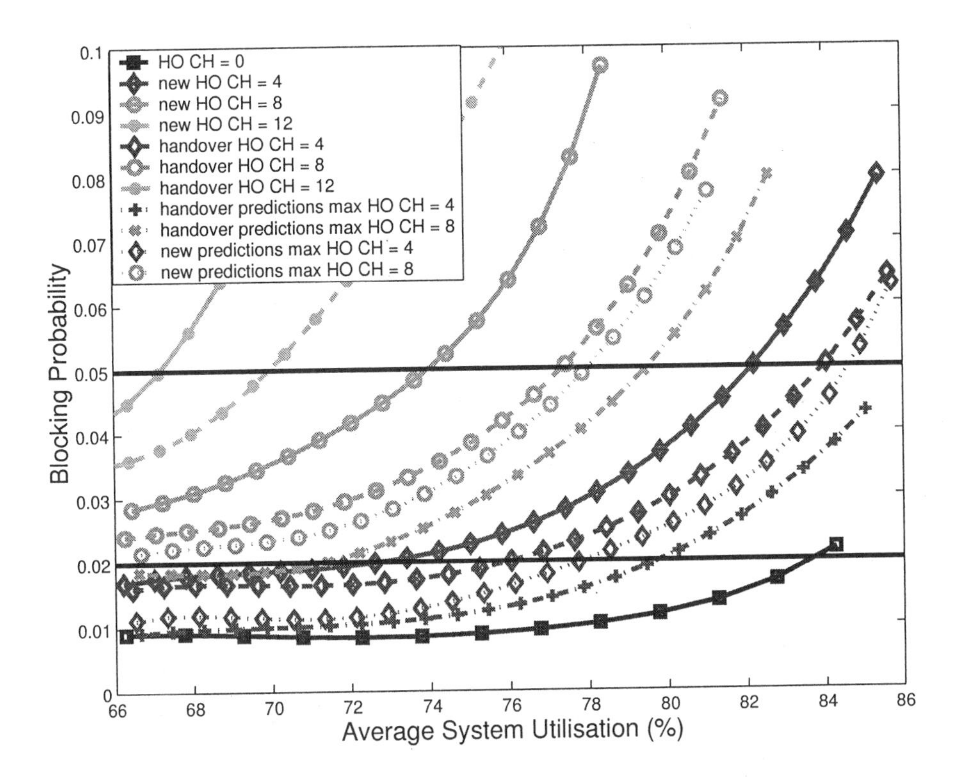

Figure 8-7. Blocking probabilities for data users with different allocation schemes.

In Fig. 8-6 and Fig. 8-7 it can be observed that with proposed algorithm very low blocking probabilities can be achieved at reasonably high utilisation. For example at 70% system utilisation handover users suffered blocking probabilities around 1% while new users suffered blocking probabilities less than 3%. These operating points are vital in third generation networks to provide GoS for service such as video conferencing. Because of the cost of spectrum, it is important to achieve the required GoS at high level of system utilisation.

6.3 Queue Delay for Data Users

Queue delay for data users is an important parameter in evaluating the performance of data users. Table 8-1 lists the mean of these delay distributions.

Table 8-1. Mean of delay distributions with different allocation strategies.

Allocation strategy	Mean delay	
	Handover users	New users
1	14.4	14.4
2	15.7	16.0
3	17.9	19.2
4	15.1	15.3
5	16.7	17.3

Allocation strategies listed in Table 8-1 are as follows:

1. No handover allocations
2. 4 handover only channels for video and voice users
3. 8 handover only channels for video and voice users
4. proposed allocation scheme (maximum number of handover only channels 4)
5. proposed allocation scheme (maximum number of handover only channels 8)

From Table 8-1 we note that the mean delay increases with increased blocking probability. This is expected as higher blocking is an indication of lack of resources in the network which results in data users being queued for longer periods. It can be observed that with the proposed allocation scheme, mean delay is reduced compared with a system with permanent allocation. This too is an indication of the effectiveness of the proposed allocation scheme, as data users spent less time in the queue compared to a system with permanent channel allocations for voice and video handover users.

6.4 System Utilisation

The aim of this research is to improve the operating utilisation of the scheme while lowering the blocking probability for new and handover users.

In Section 6.2 we discussed the effect of the proposed scheme on blocking probabilities experienced by new and handover users. In this section we present the improvements in the system utilisation that can be achieved through the proposed scheme. We define the operating utilisation of the system as the utilisation which results in less than 5% blocking probability for new users and less than 2% for handover users. Fig. 8-8 shows the operating utilisation of the system for various allocation strategies.

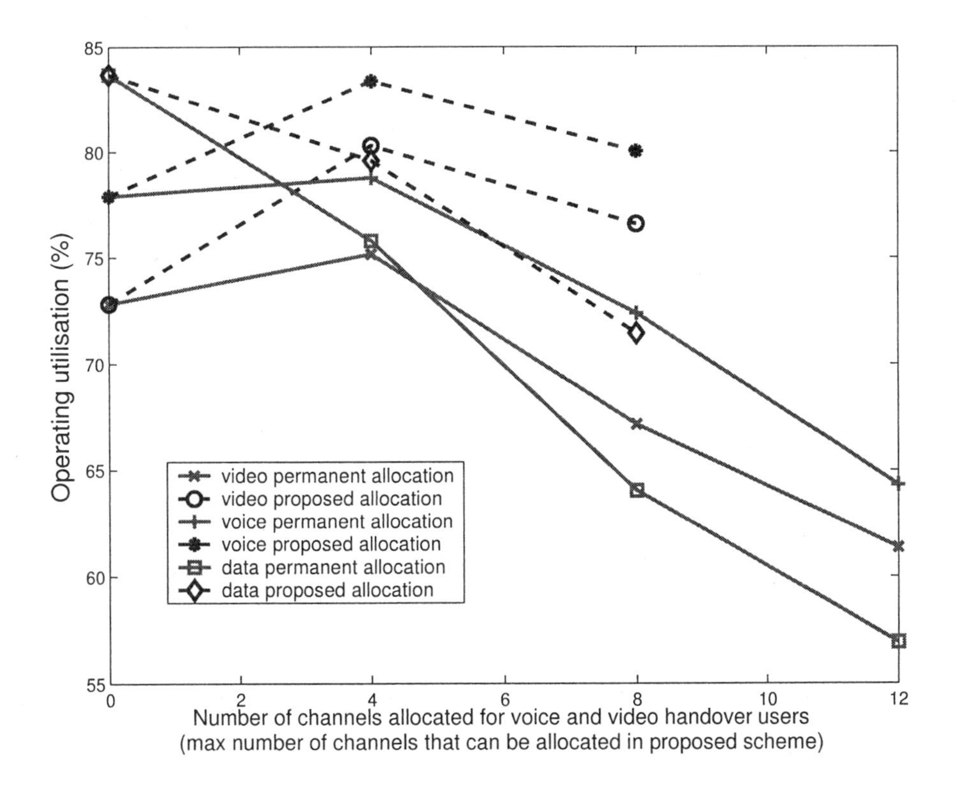

Figure 8-8. Operating utilisations with different allocation algorithms.

In Fig. 8-8, the utilisations have been determined from the blocking probabilities of the three service categories. Therefore in the legend the operating utilisation with video permanent allocation should read as "the operating utilisation which satisfies the GoS blocking probability for new and handover video users when channels are permanently allocated for voice and video handover users". The number of channels allocated to handover users has been determined from the results presented in [1] and [2]. It was shown that the optimum performance is achieved with 1 channel allocated for handover users. Therefore we have decided to allocate 4 channels for handover users. This can be broken down into 3 channels for video users and one channel for voice users. Results show that, considering video and voice, the optimum utilisation is achieved with 4 channels allocated for handover users.

Furthermore with more allocations the blocking probability for new users started to increase rapidly which led to rapid decrease in the operating utilisation. Due to resource limitation lengthy simulations for systems varying the number of handover channels between 4-8 were not performed. Therefore we cannot conclude that this is the maximum utilisation or the optimum number of handover only channels. However the results for smaller simulation models indicated that the performance of the systems with 6 and 7 channels allocated for handover users do not perform as well compared with a system with 4 handover channels. The results for these preliminary simulations are presented in [2].

Overall system utilisation for a particular allocation strategy is limited by the utilisation of the service category which demands the lowest utilisation to meet GoS for both handover and new users of that service type. Operating utilisations given in Fig. 8-8 suggest that without any handover channel allocations the overall system utilisation, considering all three service categories, is limited by the performance of video users. Although the data users can meet the GoS at 84%, the video users need the utilisation of the system to be less than 73% to satisfy the GoS blocking probabilities. Therefore we have to operate at 73% utilisation for a system without any handover allocations. The difference in the desired operating utilisation between the three user populations suggest that the utilisation of the system can be improved through a scheme which increases the maximum utilisation for video users while decreasing the maximum utilisation for data users.

Operating at 73% utilisation waste resources as there may be a way to more effectively manage the resources that are shared by video and data users. The most straight forward way of achieving this is to allocate a set of channels to be used exclusively by video and voice users where handover users always get priority over new users. With 4 channels allocated permanently for voice and video handover users, the operating utilisation for video users increases to 75% while the utilisation required to satisfy GoS for data users drops to 76%. The overall system utilisation in this case is still determined by the performance of video users. However the margin between the GoS utilisation for the three services is much smaller than that for a system without any allocation. It can also be observed that the GoS blocking probability for voice users can be satisfied at 81% utilisation which is 6% above the operating utilisation. This suggest a further prioritisation between video and voice users for improved performance. With more than 4 channels allocated for handover users, the performance starts to degrade rapidly due to the higher blocking probabilities experienced by new video and data users. Therefore the maximum utilisation that can be achieved by permanent allocations for video and voice handover users is limited to 75%.

Fig. 8-8 shows that with the proposed allocation scheme the GoS blocking probability for video users can be met at 81% utilisation, while the desired operating point for data users is 80% utilisation. GoS blocking probability for voice users can be met at 84%. These three operating points suggest the overall system utilisation is limited by the performance of data users which makes the operating utilisation of the system 80%. This is an improvement of 5% utilisation. Increasing the maximum number of channels that can be allocated for handover users beyond 4, leads to a degradation in performance. Therefore we can conclude that the maximum system utilisation is achieved with a maximum of 4 channels allocated for voice and video users using the mobility predictions and the *advance allocation algorithm*. The improvement of 5% utilisation at these operating conditions (around 75-80% system utilisation) is an important achievement. In [1] it was showed that the system blocking probabilities increased rapidly at these operating conditions, suggesting that further improvements are highly unlikely. The key achievement is that the scheme allows a system to accommodate more users while meeting their desired GoS blocking probabilities.

7. CONCLUSION

In this paper we proposed a strategy to reduce the blocking probabilities in mobile networks in this research. The proposed handover improvement scheme consisted of two major components :

1. Mobility prediction model - presented in chapter 6 of [2] and in [3].
2. Advance allocation algorithm - estimates the number of handover channels required at each cell to support the handover traffic predicted by the mobility prediction model.

The results presented in this paper showed that the proposed method considerably reduces the blocking probabilities for new voice and video users, compared to a permanent allocation scheme. This suggests that, with the proposed scheme, the new users manage to find resources more often compared with a system with permanent handover channel allocation. Therefore it can be stated that the proposed algorithm reduces the excess channels unnecessarily allocated for handover users. Furthermore the handover blocking probabilities were only increased by approximately 0.3% compared to the permanent allocation schemes. This showed that whenever a handover user request for resources the number of handover channels allocated for video and voice users are only slightly less than what is provided by a permanent allocation scheme. This strongly suggests that the proposed allocation strategy performs as expected. It allocates channels exclusively for handover users when they are required, while leaving the channels available for all users at other times, reducing unnecessary handover allocations.

Blocking probabilities for data users have different characteristics to that for voice and video users. In an over-loaded network the data users can wait in a queue for 45 seconds before being blocked by the network. Handover users can enter the queue ahead of the previously queued new users. This gives handover data users priority over new data users. With any allocation strategy the blocking probabilities for new and handover data users are increased compared to a system without any handover allocations. However with the proposed allocation schemes these increases are considerably smaller compared to the increases caused by permanent allocation schemes. This too is an indication to the effectiveness of the proposed algorithm, which shows that the data users find resources more frequently in comparison with permanent allocation schemes as a result of the reduction of unnecessary handover allocations. The results also showed that the blocking probabilities for new and handover users do not differ by the required 3% margin[1]. Therefore when the GoS blocking probability is met for handover users, the new users will be served at much smaller GoS than they require.

Operating utilisation is an important measurement in comparing the performance of different allocation strategies. Operating utilisation of the system is limited by the worst performing user category. With no handover allocation the operating utilisation of the system was 73%, which was limited by the new video users. This is expected as the video users required the highest amount of bandwidth resulting in higher chance of not being able find resources when required. With permanent allocation schemes the operating utilisation can be increased to 75%. This occurred with 4 channels reserved for voice and video handover users. Again the operating utilisation is limited by the performance of video users. With more than 4 channels allocated for handover users the performance of the system started to degrade. A maximum of 80% utilisation was achieved with the proposed advance allocation scheme based on the predicted handover traffic. Unlike the previous scenarios, under this scheme the performance of the system is limited by handover data users. The optimum performance is achieved with a maximum of 4 channels allocated for voice and video handover users. Improvement of 5% utilisation at these operating conditions is a significant achievement.

[1] The GoS blocking for new and handover users are 5% and 2%, which gives a margin of 3%

In Fig. 8-8 it was observed that the GoS blocking probability for voice users can be met at 84% utilisation, which is 4% above the selected operating utilisation. This suggests this scheme could be further improved. Under the proposed scheme we do not try to prioritise video users over voice users. As video users require more bandwidth than voice users, they should be given some advantage over voice users to find resources. The allocation algorithm can be modified to give video users a clear edge over the voice users, resulting in a drop in the blocking probability of video users. This is a possible area where we can look for further improvement in the allocation scheme.

The major drawback in using the proposed algorithm to improve the handover performance is the increased complexity of the system. All the additional mobility functionality such as the creating creation of the mobility model and making mobility predictions based on the mobility model requires a considerable amount of computer resources. This will eventually be an extra cost on the users. However it is well known that the cellular bandwidth is much more expensive than the computer resources in the current market and the processing power is becoming cheaper. Therefore we can state that the financial impact of the additional processing power is balanced by the increase in the operating utilisation that can be achieved by the proposed scheme.

REFERENCES

1. A. Jayasuriya, J. Asenstorfer, and D. Green, "Modelling Service Time Distributions in Cellular Networks Using Phase-Type service Distributions", in *Proc. Int. Conf. on Communications*, Helsinki, Finland, June 2001.
2. A. Jayasuriya, *Improved Handover Performance Through Mobility Predictions*, Ph.D. thesis, University of South Australia, 2001,
3. URL - http://www.itr.unisa.edu.au/~aruna/papers/thesis.pdf.
4. A. Jayasuriya and J. Asenstorfer, "Mobility Prediction Model for Cellular Networks Based on the Observed Traffic Patterns", in *Proc. IASTED International Conference Wireless and Optical Communications (WOC 2002),* Banff, Alberta, Canada, July 2002.
5. B. Fernandes et. al., "UMTS Task Force Report", Tech. Report, *ETSI*, Mar. 1996.
6. Special Mobile Group, "Scenarios and Considerations for the Introduction of the UMTS - draff version 2.0.0",Tech. Report, ETSI, 1996.
7. J. S. Silva, B. Barni, and B. Arroy-fernadez, "European Mobile Communications on the Move", *IEEE Communications Magazine*, pp. 60--69, Feb. 1996.
8. T. S. Rappaport, *Wireless Communications: Principles and Practice*, chapter 2,3,4, pp. 25--192, Prentice Hall, 1992.

Chapter 9

MOBILITY PREDICTION SCHEMES IN WIRELESS AD HOC NETWORKS

Mieso K. Denko
Department of Computing and Information Science, University of Guelph, Guelph, Ontario, Canada, N1G 2W1

Abstract: Mobile ad hoc networks (MANETs) are multihop networks that are capable of establishing communication in the absence of any pre-existing infrastructure. Due to frequent node mobility and unreliable wireless links, the network is characterized by unpredictable topological changes. For more robust and reliable communications, it is important that a mobile node anticipates address changes and predicts its future routes in the network. This Chapter describes prediction-based mobility management schemes for mobile ad hoc networks. We propose a Markov model-based mobility management scheme that provides an adaptive location prediction mechanism. We used simulation method to evaluate the prediction accuracy as well as the probability of making the correct predictions. The simulation results indicated that higher order Markov models have slightly greater prediction accuracy than lower order Markov models. However, the prediction accuracy decreases with increase in the probability of random movement both for network sizes and number of hops.

Key words: Ad hoc networks; location prediction; mobility management; Markov chain.

1. INTRODUCTION

A mobile ad-hoc network (MANET) is a dynamic mobile wireless network that can be formed without the need for any pre-existing infrastructure [12]. Mobile ad hoc networking has been extensively studied in recent years. However, most of these studies focused on the design of the network layer protocols such as routing (see for example [16, 20]). Node mobility brings about new challenges in the design of MANET protocols. Mobility management encompasses mechanisms with which the system tracks and maintains the location of a mobile node in a database, and manages network connections and handover. Several mobility management schemes have been proposed for cellular and wireless ATM networks in the literature. These schemes include location updating and paging, and location-prediction strategies. Location updating strategies include time-based update, movement-based update or distance-based update approaches [2]. Updates are based on a predefined threshold and performance depends on the choice of threshold values. This Chapter focuses on issues related to prediction-based location management in ad hoc networks.

Location prediction is a component of mobility management in which the system proactively tracks the future location of mobile nodes. The two main categories of location-prediction algorithms are domain-independent algorithms and domain-specific algorithms. In domain-independent algorithms, past node mobility history is used to extract context and predict the future location. The internal tables are updated in preparation to carry out future predictions. In domain-specific algorithms, the geometry of node motion and the semantics of the symbols in node's mobility history are used [3]. The accuracy of the prediction results generally depends on the mobility model and the precision of the parameter estimates.

Several prediction-based mobility management schemes are proposed for wireless communication networks (see for example [5, 10,11,18,24]). These proposals exploit the repetitive mobility pattern of mobile users and suggest prediction algorithms in order to improve the handover and reliability of the network. In [5] mobility prediction model that allows lower probability for handover and new user was proposed for wireless networks. The prediction uses current position information and system resources. In [10] a hierarchical location prediction algorithm that performs both local and global location prediction of the mobile nodes was proposed. The algorithm provides dynamic location update and virtual connection trees based on the prediction results. In [18] mobility prediction scheme for connection and handover management was proposed for cellular and wireless ATM networks. The scheme predicts the speed and direction of terminal mobility dynamically by setting up virtual connections.

Location management schemes for mobile ad hoc networks are proposed in [4,7,23]. These schemes describe how to organize and maintain location information but do not discuss how this information can be used for location prediction. The location-management scheme proposed in [4] uses location databases in order to form a virtual backbone that is dynamically distributed among the mobile nodes. Those in [7, 23] provide more scalable location-management schemes using a hierarchical addressing architecture. Recently, various mobility-prediction algorithms are proposed for ad-hoc networks

[6,15,17,21]. These algorithms differ in the parameters they predict, the information they use for prediction, and the purpose for which the prediction results are used. In [6] a link availability estimation mechanism was proposed. This mechanism predicts the probability that an active link between two nodes will be continuously available for a predicted time period. The result of this prediction was used to develop a routing metric for more reliable routing path selection. A mobility estimation and prediction scheme for cellular networks with mobile base stations is presented in [15]. The scheme uses a Robust Extended Kalman Filter (REKF) as a location heading altitude estimator for the mobile node when seeking the next base station. The purpose of the prediction is to improve connection reliability and the bandwidth of the underlying system.

A location-delay prediction scheme based on location information presented in [17] predicts locations of a mobile node based on past movement history and end-to-end delays. The location information was used in QoS routing. This mechanism assumes a piecewise linear motion pattern where nodes move only in straight lines during any two successive updates. This assumption regarding mobility patterns is not applicable in real ad-hoc network environments.

In [21], a mobility prediction algorithm is proposed for estimating the amount of time that a currently active wireless link will remain usable. GPS position information is used to predict link expiration time between two adjacent nodes. In this prediction mechanism, mobility information such speed can be obtained from GPS or sensors attached to the nodes. Nodes have their clocks synchronized using GPS clock or Network Time Protocol [13]. The prediction mechanism is extended to determine Route Expiration Time (RET) which can be applied to improve the performance of unicast and multicast routing protocols in ad hoc network. Routes that remain connected longest are used for routing to increase packet delivery ratio.

We propose an adaptive predictive mobility management scheme based on the Markov chain model. The prediction algorithm is used to determine the future location of mobile nodes. Our approach is similar to the algorithm proposed in [17] but we use the Markov chain for mobility modeling. The prediction results can be used to improve both system performance and end-user application performance. However, our main aim is to predict the location of mobile nodes for location tracking and resource reservation. Our model can predict a single location as well as a sequence of locations. The scheme uses information on node movement history, with the level of history depending on the order of the Markov chain model applied.

The rest of this Chapter is organized as follows: Section 2 describes the Markov chain; Section 3 describes the system model; Section 4 describes the parameter-estimation and location-prediction algorithms; Section 5 provides a summary and describes future research work.

2. THE MARKOV CHAIN MODEL

A discrete time stochastic process $\{X_n : n = 0,1,2,...\}$ with discrete state space $S_n = (0,1,2,3,...,n)$ is a Markov chain [14] if,

$$P(X_n = s_n \mid X_0 = s_0, X_1 = s_1, ..., X_{n-1} = s_{n-1})$$

$$\Rightarrow P(X_n = s_n \mid X_{n-1} = S_{n-1}) \tag{9.1}$$

The Markov chain can be represented by state transition diagram as shown in Fig. 9-1.

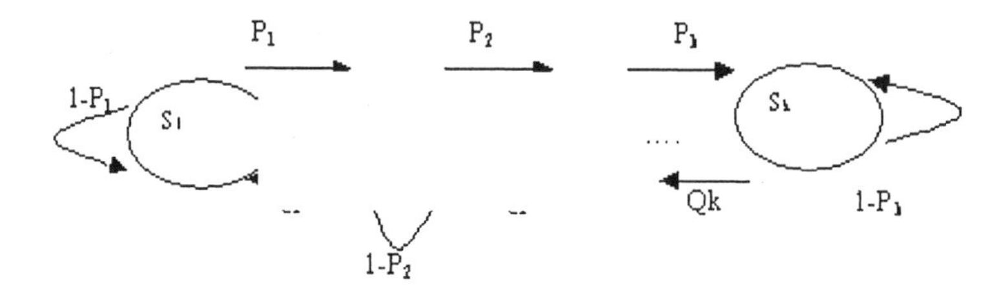

Figure 9-1. The State transition diagram

In a steady-state condition, the mean number of outgoing paths are the same as the mean number of incoming paths. Let P_{ij} be the transition matrix, the steady-state probability (π_i) is determined by solving equations (9.2) and (9.3).

$$\sum_{i=0}^{\infty} \pi_i P_{ij} = \pi_j \tag{9.2}$$

$$\sum_{i=0}^{\infty} \pi_i = 1 \tag{9.3}$$

For the first-order Markov chain model, the next state depends only on the current state. If a_{ij} is the transition probability from state i at time k to state j at time k+1, then the transition probability matrix P of the Markov chain can be represented as shown by the following equation:

$$P = \begin{bmatrix} a_{00} & a_{01} & \cdots & a_{0n} \\ a_{10} & a_{11} & \cdots & a_{1n} \\ \vdots & \vdots & \ddots & \vdots \\ a_{n0} & a_{n1} & \cdots & a_{nn} \end{bmatrix} \quad (9.4)$$

The conditional probability of transitions to an nth state given that $k=n-1$ previous states have occurred, gives the kth order Markov chain and is computed as follows:

$$p(X_n \mid X_{n-1},\ldots X_{n-k}) = p(X_n = s_n \mid X_{n-1},\ldots,X_{n-k}) \quad (9.5)$$

For a sequence S of length L, the probability of the observed sequence for the first-order Markov chain is computed as follows.

$$P(S) = p(s_L \mid s_{L-1},..,s_1)p(s_{L-1} \mid s_{L-2},..,s_1)...p(s_1)$$

$$= p(s_L \mid s_{L-1})p(s_{L-1} \mid s_{L-2})...p(s_1)$$

$$= p(s_1)\prod_{i=2}^{L} p(s_i \mid s_{i-1}) \quad (9.6)$$

3. THE SYSTEM MODEL

User movement has some degree of regular and irregular patterns. However, the node's current and past mobility pattern can provide useful information that can be used to anticipate the node's future location. Neither a purely random mobility model such as the memory-less random walk model nor a deterministic model such as a linear motion is suitable for modeling node mobility in MANETs. A more realistic mobility model for node mobility should include both random and regular components although the extent of regular and irregular contents may vary from one application to another. Thus the movement of a mobile node is modeled using a discrete-time Markov model. A discrete-time Markov chain model can be described by a tuple (S_i, P_{ij}, Q_i) where S is the state space, P_{ij} is the transition probability matrix and Q_i is the initial probability distribution.

The Markov chain model allows probabilistic predictions of the future locations given the previously visited locations. It maintains the transition matrix, initial state probability distribution, and states gathered at the nodes. Each row of the transition matrix represents the transition flow from state i to state j while column represents the transition flow into state j. To set up a Markov model, the state space (S), the transition probability (P_{ij}) and the initial state distribution (Q_i) need to be determined. The algorithm for setting up the Markov chain determines the initial probability vector, decides the order of the Markov chain and computes the transition probabilities.

3.1 Description of the model and states

We model the mobile ad-hoc networks probabilistically as a discrete-time Markov chain process with finite states where states correspond to location grids. The nodes' movement in the locations grid can be seen as stochastic process which has S as the state space. In our model, the physical area of network coverage is divided into a grid of size $a \times b$ (where a and b are dimensions of the grid) similar to the proposal in [9]. A grid forms a location area within which all mobile nodes can directly communicate. We use the terms grid and location interchangeably to refer to the location of a mobile node.

Each node has a unique ID and can move within the location grid on a grid-by-grid basis. A grid can contain multiple nodes and the number of nodes in the grid depends on the communication range and the grid size. Larger location grids and radio communication ranges allow more nodes to communicate, when compared to smaller grids and ranges.

While the grid ID is used to identify each node's current state, within a grid each node's current location is identified by the entries (i,j) where i and j represent the location coordinates. When a node moves from one grid to another grid its state changes in correspondence with the current grid ID. Each mobile node is equipped with a position tracking device, such as GPS for outdoor use or infrared sensing devices for indoor use depending on the application environment. Node mobility parameters such as the current location coordinates, speed and direction of motion are provided by GPS clock or using the Network Time Protocol proposed in [13].

3.2 State transition and data storage

We assume that the probability of visiting a location in the future depends on a fixed number of previously visited locations. State transitions can occur due to node movement from one location grid to another location grid, or due to changes in mobility rate or network resource requirements. We consider only state transitions related to a node's position in the location grid. The state of a node is the grid ID of the current location grid, and the state changes when a node enters a new location grid. At every time interval, a node remains in the same state or moves to another state and possibly returns back to the same state according to the transition probability associated with the state. The transition of the mobile node from one state to the next state evolves according to the transition probabilities.

In our model, each node builds its own Markov model where the necessary state information is maintained. Each mobile node collects and maintains: (a) a table that stores the node ID, grid ID, position information obtained from GPS and dwell time in the location grid; (b) the transition probability matrix; and (c) the state sequence, current state and predicted location grid. A node also maintains additional information such as its current neighbors. Transition probabilities are constructed from the previous movement history data maintained at each node. To reduce storage overhead at mobile nodes, the sparse transitional probability matrix can be compressed using the algorithm proposed in [19]. This compression algorithm reduces the sparse transition matrix into a smaller matrix and provides reasonably accurate estimations of the transient behavior of the Markov model.

Location prediction has several advantages when compared to traditional location-update and paging-based strategies. First, it proactively estimates future locations of communicating nodes and minimizes location updating and paging costs, thereby reducing communication latency. In location tracking applications, in the case of location miss, search can begin at the predicted location. Second, it improves routing performance by enabling fast route discovery and maintenance mechanisms. Thus topology can be reconstructed and routing information can be updated before packet forwarding. Third, it allows resource pre-allocation before the mobile node moves to the new location. For resource pre-assignment, location will be made along the paths with better resources at or in the neighborhood of the predicted location. However, the efficiency of such allocation depends on the accuracy of the prediction model. Fourth, it optimizes data transmission by allowing the selection of locations with better resources. Fifth, it improves handover management, resource reservation and call-admission control mechanisms.

4. PARAMETER ESTIMATION AND PREDICTION

4.1 Parameter estimation

The Markov chain model parameters are estimated using the maximum-likelihood technique. Consider a Markov chain model of a random sequence states space S. The maximum-likelihood function $f(P,Q|X)$ with unknown parameters, P and Q is given as follows when n is number of states:

$$L=f(P,Q|X)=\prod_{i=0}^{n} Q_i^{n(i)} \prod_{ij=0}^{n} P_{ij}^{n(i,j)},\ i,j\in S \tag{9.7}$$

The value $n(i,j)$ is the number of times the state transition from i to j has occurred in the observed sequence, and $n(i)$ is number of times the initial state is equal to i. Given sequences of states, the maximum likelihood estimates of the transition probabilities and the initial probabilities are computed as follows.

$$P=[P_{ij}]=\frac{n(i,j)}{\sum_{j'} n(i,j')} \tag{9.8}$$

$$Q_i=\frac{n(i)}{\sum_{i'} n(i')} \tag{9.9}$$

An element in the transition matrix P_{ij} can be interpreted as the probability of transition from state i to state j in one step. The parameter estimates for the higher-order Markov chain are interpreted as the probability of a node moving from one location to another location, given a specified sequence of previous states. The estimates of the parameters are determined from finite runs of Markov chain model. If the initial probability distribution of the states denoted by a row vector (Q_0) is known, the probability distribution of the next state in the Markov chain (first-order) is computed as follows for any k:

$$Q_k = Q_0 P^k \tag{9.10}$$

4.2 Location and path prediction

We built a Markov model for location prediction based on a sequence of past movements of mobile nodes. The transition probability matrix of the first-order and higher-order Markov chain is computed using current and previous states depending on the order of the Markov model. The model is then used for prediction by computing the conditional probabilities of visiting other locations in the future given the current location or a sequence of previously visited locations.

Mobility prediction algorithms can predict a single location or a sequence of locations. The location prediction can provide vital information for applications such as location management (updates or paging), route discovery (for on-demand routing protocols) or resource management (resource pre-allocation and reservations).

Based on a Markov model of different orders, we proposed three types of prediction algorithms:

1. *One-step prediction:* This algorithm is used to perform a single-step prediction into the future.
2. *Multi-step prediction:* This algorithm is used to perform a k-step prediction steps into the future for any $k>1$.
3. *Path sequence prediction*: This is used to predict a sequence of future locations for each mobile node. The algorithm uses a sequence of states with the highest average transition probability.

The prediction accuracy is important because inaccurate prediction results in poor system performance or end-user applications. For example, in resource pre-allocation, poor prediction results in waste of resources or inefficient resource utilization. Similarly, inaccurate prediction may cause routing and location tracking overhead as the cost of route maintenance or location update increases with increase in the difference between predicted and actual location of the mobile node. For a single location prediction, the prediction error is the difference between the predicted location and the actual location. The accuracy of the predicted path sequence is computed as the fraction of the predicted sequence (model estimate) that matches the actual paths. We examined the performance of the models by investigating the probability that they make correct predictions. For each algorithm, the following were investigated:

1. *The Probability of correct prediction (PCP).* This is the probability that the location grid visited is the one predicted by the model as the most likely location. PCP is computed as the proportion of visits to a location over a period of time. This gives an overall accuracy of the correct prediction.
2. *The conditional probability of correct prediction (CPCP).* This is the probability that the location grid (X_n)is visited given the past visits, based on the fact that there is a higher conditional probability on those paths.

Since X_n is a random variable, it might be important to investigate the effect of the uncertainty of this random variable on the accuracy of the Markov models. This requires performing entropy analysis and investigating the prediction stability of the Markov models over time. However, this was not considered in this study.

5. PERFORMANCE EVALUATION

In our simulation, we generated mobility patterns as a sequence of finite states for a specified time period denoted as (t,s_t) where t is the unit of time, for $t = (0,1,2,3,...,n)$ and s_t is the state at time t. The system maintains node mobility history for each node. Node movement can be random or regular based on the probability of random movement specified at the beginning of the simulation. We built the Markov models and performed predictions. We computed the prediction accuracy and the probability of making correct predictions for various parameter settings.

5.1 The effect of Markov model prediction steps and number of predictions

In this section we investigated the effect of the order of Markov model, number of predictions made and prediction steps on the prediction accuracy. The prediction step refers to the length of prediction time unit in the future. Number of prediction refers to the number of predictions made. The purpose of this metric is to investigate the accuracy attained when several predictions are made (instead of a single guess) and select the best predictions.

The performance of each prediction algorithm was evaluated by estimating the prediction accuracy. The accuracy of each prediction algorithm was computed using a grid method and path method. In the grid method, an absolute difference between the predicted location grid and the actual location grid in which the node is located was computed. In the path method, the difference between predicted and observed path sequences was computed. For this purpose, a specified length of sequences was used.

To evaluate the performance of our prediction algorithms, we simulated them using the NS-2 simulation environment [22]. The aim of the simulation experiment was to investigate the effect of various parameter settings on the accuracy of the proposed algorithms. Fig. 9-2 and Fig. 9-3 show the prediction accuracy as the function of the number of prediction steps and the order of the Markov model respectively. The results in Fig. 9-2 indicate that the precision accuracy decreases as the prediction step increases for all the simulated models. However, the higher-order Markov model has better overall prediction accuracy. The results in Fig. 9-3 indicate that prediction accuracy is slightly lower for longer prediction steps than for shorter ones. The results imply that it is possible to accurately predict the future location of a mobile node with relatively smaller prediction steps and higher-order Markov models.

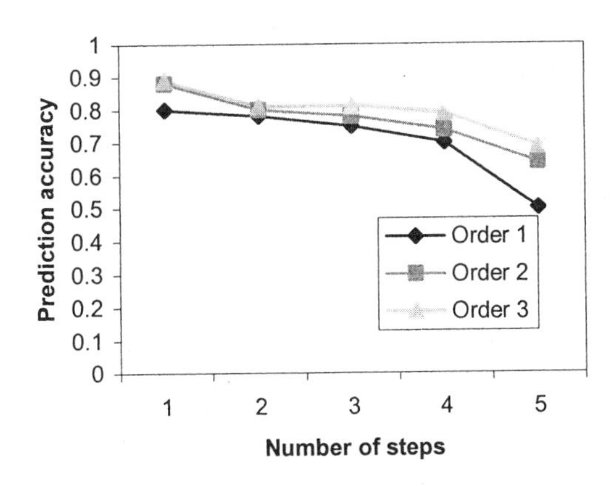

Figure 9-2. Accuracy (Markov models).

Fig. 9-4 and Fig. 9-5 show the probability of making the correct prediction as a function of the order of Markov model and the number of predictions respectively. The results in Fig. 9-4 indicate that the conditional probability of making the correct prediction increases with the order of Markov model. This implies that longer node-movement history results in better predictions. The results in Fig. 9-5 indicate that the probability of making a correct prediction increases both with an increase in number of repeated predictions made, and in the order of Markov model. Thus prediction capability of the algorithms increase as the function of the number of predictions made.

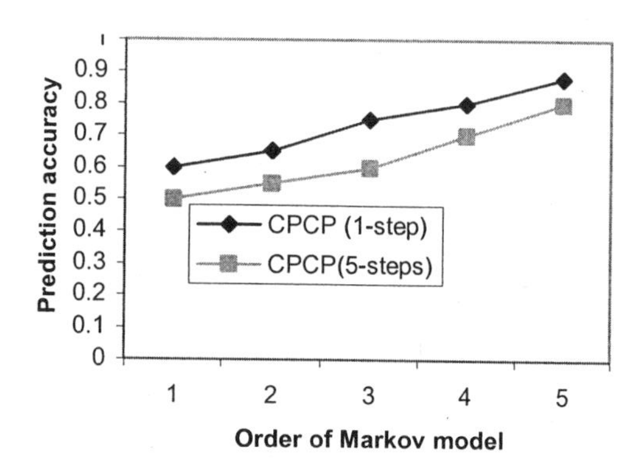

Figure 9-3. Accuracy (prediction steps).

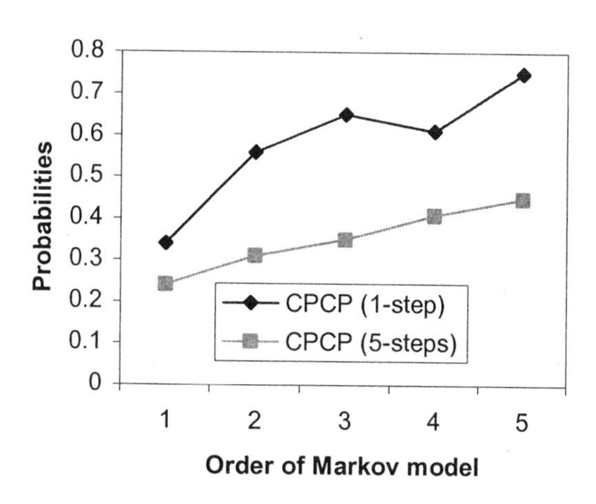

Figure 9-4. Probability of making a correct prediction (prediction steps).

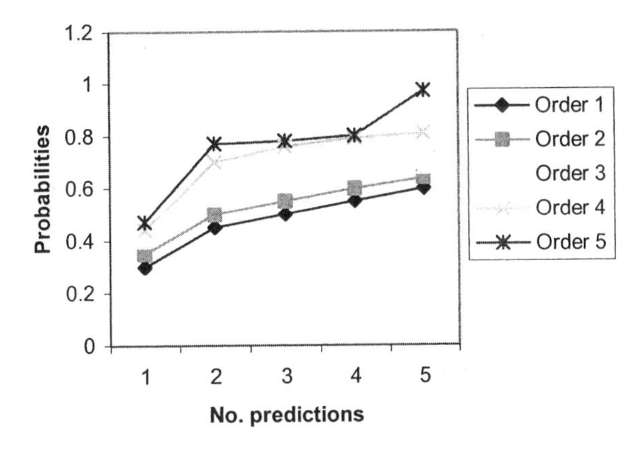

Figure 9-5. Probability of making a correct prediction (Markov models).

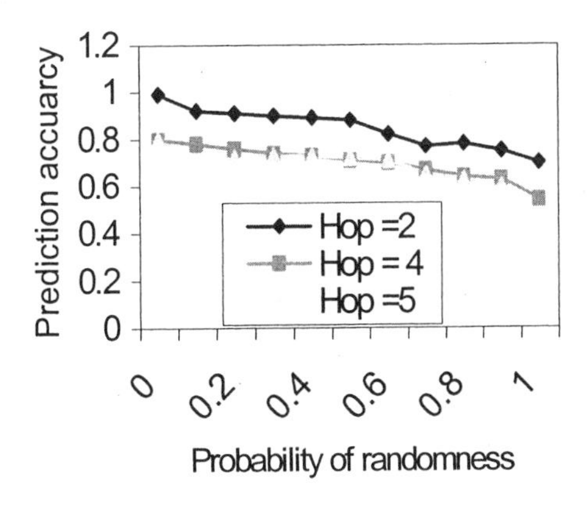

Figure 9-6. Probability of randomness v.s. number of hops.

5.2 The effect of network size and hop length

In this section the simulation experiment was carried out to investigate the effect of random movement on the accuracy of the prediction results using the second order Markov chain model. Fig. 9-6 and Fig. 9-7 show the effect of randomness under three different hop lengths and network sizes respectively. Both results indicate that there is a decline in prediction accuracy with increase in the network size and number of hops. The decline in prediction accuracy with increase in network size is due to the fact that larger networks result in longer hops. These results imply that it is possible to accurately predict the future location of a mobile node with relatively lower degree of randomness and smaller number of hops.

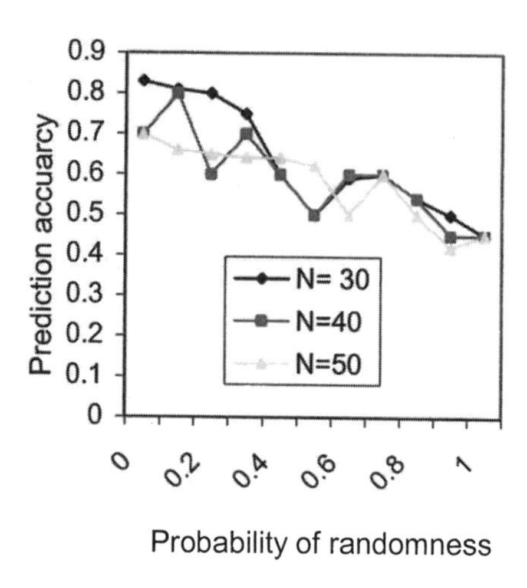

Figure 9-7. Probability of randomness v.s. network size.

The application of the current prediction results is limited to location tracking which involves both single location prediction and a sequence of locations (paths) prediction. In the future, we intend to use the prediction results in other applications (end-user or system-level) and evaluate their performances. Among these are: (a) Evaluating the efficiency of resource utilization in resource management; (b) Evaluating the performance gain in route discovery and maintenance in network routing; and (c) Investigating the extent of reduction in the cost of location updates in location management tasks.

6. SUMMARY AND FUTURE WORK

Due to limited network resources and random, frequent topological changes, mobility management techniques developed for infrastructure-based networks cannot be directly applied to mobile ad-hoc networks. We proposed a predictive mobility management scheme using the Markov chain model. The Markov model provides detailed computations of the probability distributions used to predict the future location of mobile nodes. This mobility prediction scheme can provide several functions such as service and data pre-connection, and resource pre-assignment at the future locations of the mobile nodes.

The simulation results indicated that the prediction is reasonably accurate for lower prediction steps and higher-order Markov models. Furthermore, the simulation results based on network size and number of hops indicated that better prediction can be achieved for smaller network sizes and hop counts at lower probability of randomness. We are currently working on further performance evaluation using additional metrics such as computational complexity and storage requirements and larger network sizes. We intend to investigate prediction stability for each model using entropy analysis. The accuracy of the prediction results will be further analyzed using statistical techniques. Our future work will also include investigating node mobility along an arbitrary topology using the multidimensional Markov chain model and developing hybrid mobility management schemes.

REFERENCES

1. I.F. Akyildiz, J.S.M. Ho and Y.B. Lin, "Movement-based location update and selective paging for PCS networks," *IEEE/ACM Transactions on Networking*, 4(4), pp. 629-638, August 1996.
2. A.Bar-Noy, I. Kessler and M. Sidi, "Mobile Users: To update or not to update? " *IEEE/ACM Transactions on Networking*, Vol. 4, pp. 629-638, August 1996.
3. C. Cheng, R. Jain & E. van Den Berg, "Location prediction algorithms for mobile wireless systems," M. Illyas and B. Furht, eds. *Handbook of Wireless Internet*, CRC Press, 2003.
4. Z. Haas and B. Liang, "Ad hoc mobility management with uniform quorum systems," *IEEE/ACM Transactions on Networking*, 7(2), pp. 228-240, April 1999.
5. A. Jayasuriya and J. Asenstorfer, "Mobility Prediction Model for Cellular Networks Based on Observed Traffic Patterns," *Proceedings of the IASTED International Conference, Wireless and Optical Communication* (*WOC2002*), pp. 386-391, Banff, Canada, July 2002.
6. S. Jiang, D. He & J. Rao, "A prediction-based link availability estimation for mobile ad hoc networks," *INFOCOM 2001*.
7. K. K. Kasera and R. Ramanathan, "A location management protocol for hierarchically organized multihop mobile wireless networks," *Proceedings of the IEEE ICUPC*, San Diego, 1999.
8. B. Liang and Z. Haas, "Predictive distance-based mobility management for PCS networks," *INFOCOM* 1999, pp. 1377-1384, March 1999.

9. W. H. Liao, Y. C. Tseng & J. P. Sheu. GRID, "A fully location-aware routing protocol for mobile ad hoc networks," *Technical Report*, National Central University, 2001.
10. T. Liu, P. Bahl and I. Chlamtac, "Mobility modeling, location tracking and trajectory prediction in wireless ATM networks," *IEEE Journal of Selected Areas in Communications*, Vol. 16, pp. 922-936, August 2001.
11. G. Liu and G. Maguire, "A class of motion prediction algorithms for wireless computing and communications," *ACM/Baltzer MONET*, 1(2), pp. 113-121, 1996.
12. J. P. Macker and M.S. Corson, "Mobile ad hoc networking and IETF," *ACM Mobile Computing and Communications Reviews*, 2(1), pp. 9-14, January 1998.
13. D.L. Mills, Internet Time Synchronization, "The NetworkTime Protocol," *IEEE Transactions on Communications*, 39(10), pp. 1482-1497, Oct 1991.
14. A. Papoulis, *Probability, random variables and stochastic processes*. 3rd ed., McGraw-Hill, 1991.
15. P.N. Pathirana, A.V. Savkin & S.K. Jha, "Mobility modelling and trajectory prediction for cellular networks with mobile base stations," *MobiHoc 2003*, pp. 213-221.
16. E.M. Royer and C. K. Toh, "A review of current routing protocols for ad-hoc mobile wireless networks," *IEEE Personal Communications Magazine*, pp. 46-55, 1999.
17. S. H. Shah, K. Nahrstedt, "Predictive Location-Based QoS Routing in Mobile Ad Hoc Networks," *Proceedings of IEEE International Conference on Communications (ICC 2002)*, New York, NY, April 28th - May 2nd, 2002.
18. N. Shenoy, A. Dadej, J. Asenstorfer, P.Alexander, "Mobility Prediction for Optimal Handover and Connection Management in Mobile Multimedia," *Proceedings of the Third Asia Pacific Conference in Communication '97,* Sydney 7-10 December, pp. 1236-1240.
19. W.M. Spears, "A Compression Algorithm for Probability Transition Matrices," *SIAM Matrix Analysis and Applications*, 20(1), pp.60-67, 1998.
20. I. Stojmenovic, "Position based routing in ad hoc networks," *IEEE Communications Magazine*, 40(7):128-134, July 2002.
21. W. Su, S.J. Lee & M. Gerla, "Mobility prediction in wireless networks," *Proceedings of IEEE MILCOM 2000*, Los Angeles, CA, Oct. 2000.
22. *The NS-2 Network Simulator*. http://www.isi.edu/nsnam/ns/
23. Xue, B. Li, and K. Nahrstedt, "A scalable location management scheme in mobile ad-hoc networks," *Proceedings of IEEE Conference on Local Computer Networks, LCN*, 2001.
24. F. Yu and V. Leung, "Mobility-based predictive call admission control and bandwidth reservation in wireless cellular systems," *Computer Networks*, Vol.38, pp. 577-589, April 2002.

PART 3:

PERFORMANCE OF ADVANCED NETWORKS AND PROTOCOLS

Chapter 10

AN OVERVIEW OF STREAMED DATA AUTHENTICATION TECHNIQUES

Beata J Wysocki
School of Electrical, Computer, and Telecommunications Engineering, University of Wollongong, NSW 2522, Australia

Abstract: In the chapter we present some of the techniques proposed for authenticating digital streams. We start with the so-called block signatures, where every packet contains full information needed for its authentication. Then, we describe two techniques based on a simple hash chain for an off-line authentication and its modification where one-time signatures are used for an on-line authentication. Later, we consider the TESLA scheme for authenticating streams in case of synchronous systems. This is followed by augmented hash chains with signature packets used once per a sequence of packets rather than for the whole stream. The chapter is concluded with a short comparison of the considered techniques.

Key words: Streamed Data, Authentication, Hash Chain, Digital Signature, Gilbert-Elliot Channel

1. INTRODUCTION

Streamed data or a stream of data packets is generated by a specialized application on a server machine, and then sent in a form of autonomous packets over the Internet (or other packet switched network). The process of splitting a large block of information into individual packets by a server application needs to be distinguished from breaking a large file into packets at the transport layer (e.g. TCP [1]) before sending them through the packet switched network.

In case of streaming, a file is divided into packets at the application layer. These packets are of the size that would fit the size of a packet in the underlying packet network. In the case of Internet, it is usually not more than 1000-1500 bytes [2]. As soon as a packet arrives at the transport layer of the receiving host, it is passed to the application layer. The client application can then execute or play the received packet as it comes. It is the client's application's task to discard packets arriving out of order or being corrupted.

The streaming involves mainly real time applications [3] (e.g. Real Player); hence, it usually uses the UDP [1] transport, which means that no retransmissions or integrity checks are performed at the transport layer. These must be catered for by the application itself. The advantage of UDP over TCP is its speed and lower communication overhead but there is no guarantee for packet delivery or for its correctness. In addition, the client's machine utilizes the received data at more or less the input rate, which means that it cannot buffer large amount of the received data [4].

The importance of the need to authenticate streamed data manifests itself in a fact that the recipient would like to have an option to verify the source and authenticity of the received information. For example, it is important to the listeners of an Internet radio station that the audio stream they receive was really broadcast by the station they listen to. On the other hand, it is equally important to that station that only the content it broadcasts is attributed to it. Malicious parties should be prevented from injecting commercials or offensive material into the stream, which might be quite easy in the case of wireless networks. Moreover, those broadcasts must be non-repudiable, which means that a positive verification has to be enough to hold the transmitter responsible for the content. An answer to these problems is the use of digital signatures. However, the digital signature technology has been developed for single messages, not for continuous streams of autonomous packets [5-7].

Authentication of a single message or a file transferred from a server to a remote client can be done using one of the standard schemes, like digital signature. The signature or other authentication features are generated by the application and appended in some form to the file, which is then passed to the transport layer, where it is packetized. The part that contains the signature is not treated in any way different to the other contents of the message (file). At the remote client machine, the transport layer reassembles the received packets, checks the message (file) integrity and passes the whole message (file) to the application layer, where verification of the message (file) is performed.

For the streamed data, this approach is not feasible. The application layer at the server generates autonomous packets, and the transport layer considers these packets as separate messages. At the remote client's machine, the packets are played out as they arrive. This means that they are not assembled back at the client's application layer. In the optimal solution, they are not even buffered. Hence, it is not possible to directly apply an authentication approach used for non-streamed data to the streamed data. The nature of a digital stream creates the need to authenticate and verify each of the received packets. As a result, each packet must carry some features verifiable by a remote client.

Several authentication/verification schemes have been proposed in literature, e.g. [5-7,9-11]. Most of them can be classified into two major groups:

- The schemes where every packet carries the full information necessary to verify the packet [6,7], which in some sense resembles signing of every packet individually.
- The schemes where only one packet in the stream (usually the first or last) is signed and the other packets are connected with the signature by a chain of hashes [5,9,11].

There are, of course, several modifications to the second group, e.g. [11]. However, all of the schemes mentioned above suffer from the fact that once the verification chain is broken, then there is no means to verify packets incoming after the break.

The way to avoid this drawback and somehow combine the benefits of both groups has been proposed by Golle and Modadugu in [10]. They proposed to divide the stream into sequences of N packets, with each of the sequence being individually verifiable. Hence, a break in the verification chain would result in rejection of a maximum of one sequence of packets but not all packets coming after the break. The technique proposed in [10] is resistant to packet losses with bursts of up to a predefined length. However, even an isolated single packet loss, when the packet containing a signature is involved, results in a loss of the whole sequence of packets. This has been later modified by Wysocki et al in [12], where they proposed a technique immune to the loss of a signature packet at the expense of possible longer verification delays if the signature packet is lost.

The chapter is organized as follows. In Section 2, we briefly introduce techniques proposed to authenticate digital streams based on the so called 'block signatures' [6,7]. Then, in Section 3, we will describe techniques employing 'verification chains'. Section 4 considers the TESLA [9] scheme introduced for systems where time synchronization can be maintained. The techniques using graph based approached are described in Section 5, with an example of Golle and Modadugu method [10] and a way to protect such schemes against the loss of signed packets [12]. The chapter is concluded in Section 6 with the performance comparison of the described authentication schemes.

2. BLOCK SIGNATURES

Signing every packet of the data stream can be considered as a direct extension of the message signing technique into authentication of digital streams. This approach, however, involves unacceptable overheads, both computational and communication ones. Some reduction in those overheads without compromising the benefits of having every packet independently verifiable has been proposed in [6].

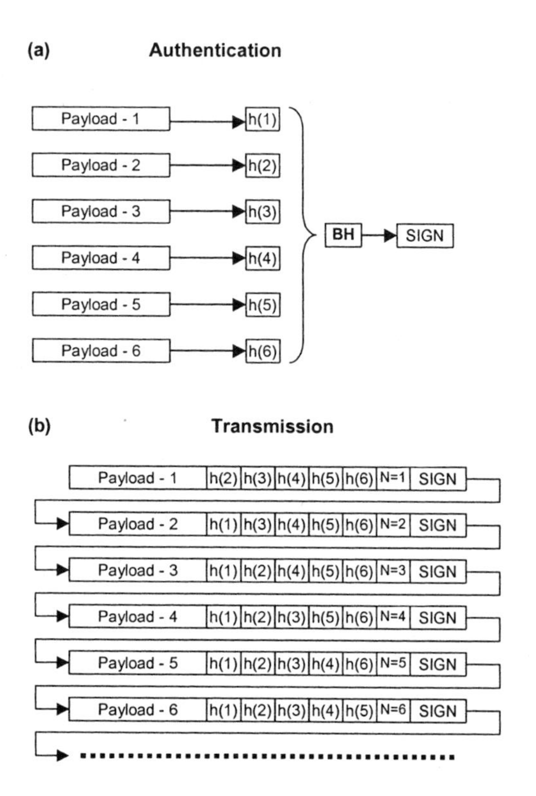

Figure 10-1. The star chaining technique; $h(i)$, $i = 1, \ldots, 6$, denote hashes for individual packets, N is a position of a packet in the block, $N = 1, \ldots, 6$, *SIGN* is a standard digital signature for the block hash *BH*; (a) an authentication process, (b) transmission.

In [6], Wong and Lam proposed a star chaining technique for signing digital streams. During the authentication phase, a block of 'm' packets is formed and hash functions are calculated for every packet in a block, and then for a sequence of these individual hashes a block hash is calculated.

After that, the block hash is signed using a standard digital signature algorithm. For packets to be individually verifiable, each packet needs the full authentication information, called a packet signature. The packet signature consists of the block signature, the packet number in the block, and the hashes of all other packets in the block. The authentication process and the transmission of the resulting packet block are illustrated in Fig. 10-1 [12].

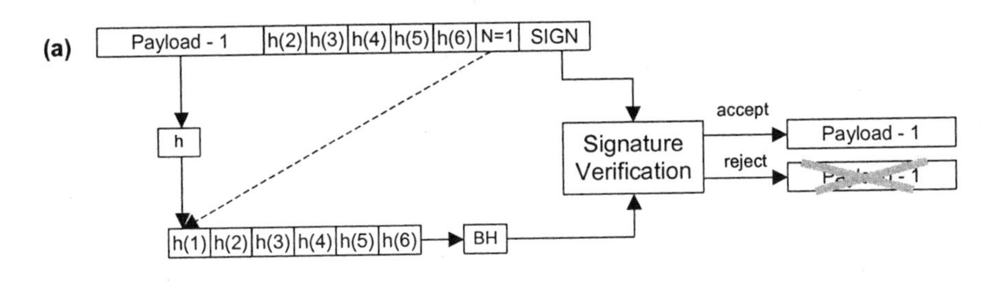

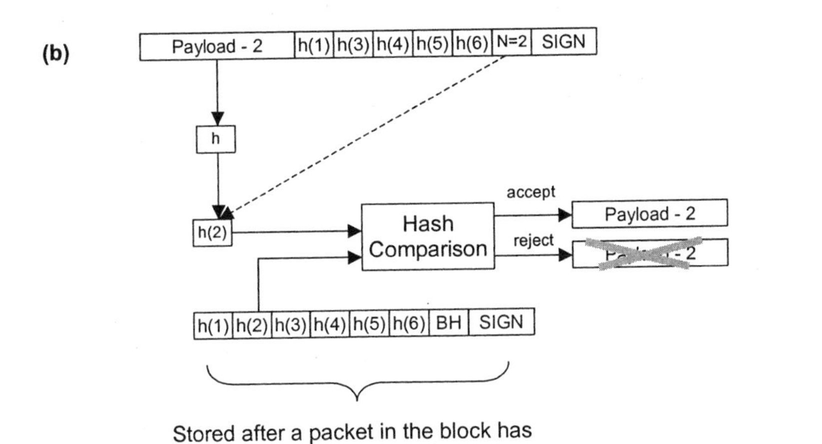

Figure 10-2. Verification process for the star chaining technique; (a) verification of the first packet in a block, (b) verification of the second and other packets in the block.

The verification procedure is as follows:

- For the first packet received from the block, the verifier calculates the packet's hash.
- Based on the calculated packet hash, the packet number, and the hashes of other packets in the block contained in the packet signature, the verifier computes the block hash.
- Using the calculated block hash, receiver verifies the block signature; if it is correct, the packet is accepted, if not it is rejected.

For all other packets from the block, the verifier needs only to calculate the new packet hash and compare it to the hash contained in the packet from this block previously positively verified. If they agree, the new packet is verified. The verification process is illustrated by Fig. 10-2.

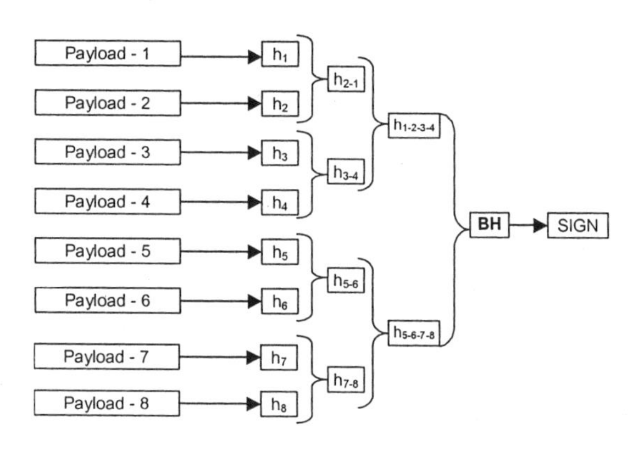

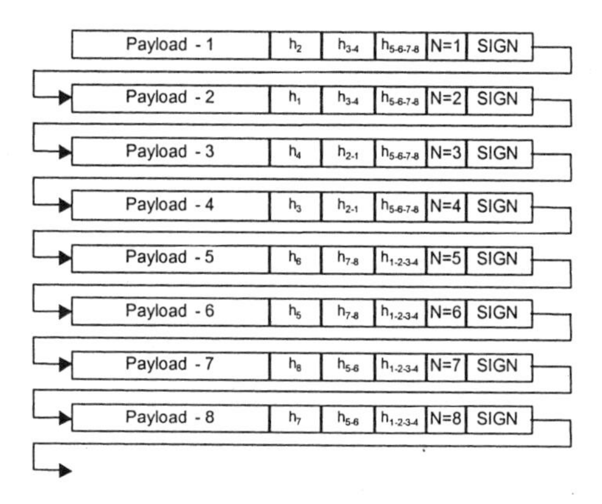

Figure 10-3. Tree chain authentication process and transmission of the resulting packet block; h_i - hashes of individual packets, $h_{i\text{-}j}$ - hashes computed based on the hashes h_i and h_j, $h_{i\text{-}j\text{-}k\text{-}l}$ – hashes computed based on hashes $h_{i\text{-}j}$ and $h_{k\text{-}l}$, $i,j,k,l = 1, \ldots, 8$.

A generalization of the star chaining technique is also proposed by Wong and Lam in [6]. In this generalized scheme, the block hash is computed as a root node of an authentication tree, (see Fig. 10-3). In such a tree, the packets' hashes are the leaf nodes of the second-degree authentication tree, with other nodes of the tree computed as hashes of their children. For example, in Fig. 10-3, the parent of leaves D_3 and D_4 is $h_{3\text{-}4} = h(D_3, D_4)$. The block signature is calculated on the block hash.

In tree chaining, a packet signature (packet overhead) consists of:

- The block signature.
- The packet number in a block.
- Siblings of each node in the packet's path to the root.

From Fig. 10-3, one can notice that the communication overhead can be reduced compared to the star chaining scheme. However, this is paid by the increase in computational overhead.

During the verification process the packet's path to the root is verified, i.e. all nodes on that path. The procedure is similar to that for the star-chaining scheme and is illustrated in Fig. 10-4.

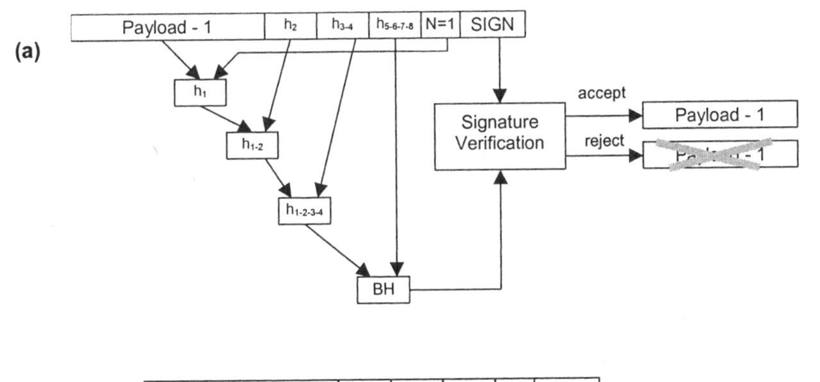

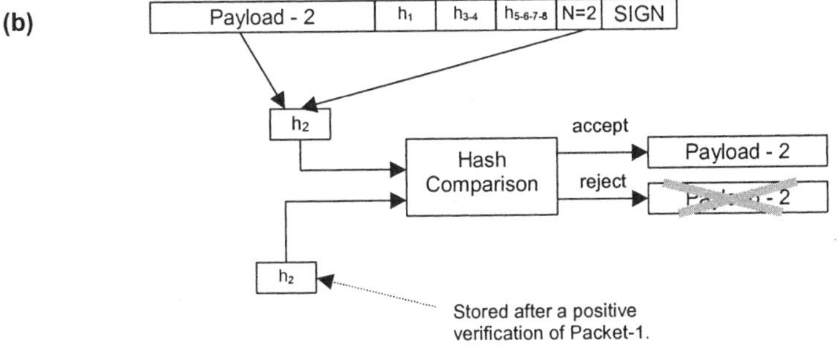

Figure 10-4. Verification process for the tree authentication chain; (a) verification of the first packet in the chain, (b) simplified verification of the second packet.

The nodes, once positively verified, can be later used for verification of other packets' paths in the block. This is illustrated in Fig. 10-4 where it is shown that verification of the Packet - 2, after the positive verification of the Packet-1, requires calculation of just one hash, and comparison of it with the h_2 stored from the Packet-1.

Verification process, however, is not always as simple as presented in Fig. 10-4. Since all nodes on the path to the signature must be verified, 3 hashes must be computed when the completely new path to the signature has to be verified . We show such a case in Fig. 10-5, where the verification of Packet - 5 is presented.

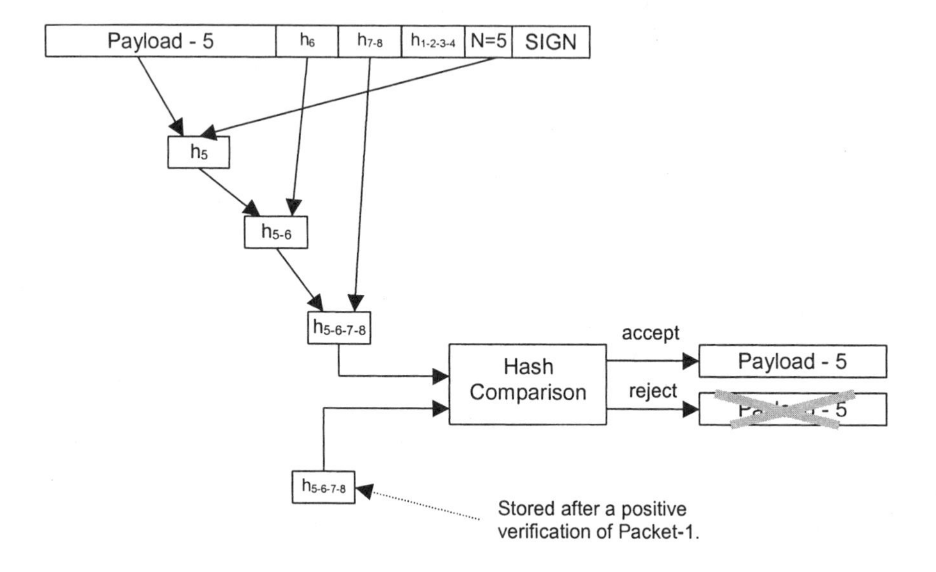

Figure 10-5. Verification of Packet- 5 in the tree authentication chain.

The main advantage of both schemes is their ability to independently verify each of the received packets. Hence, packets lost or tampered with can be discarded independently. On the other hand, both methods involve high computational overhead and quite high communication overhead [10].

3. HASH CHAINS

3.1 Hash Chain of Gennaro and Rohatgi

The simplest scheme using a hash chain was proposed by Gennaro and Rohatgi in [5]. It involved the use of just one full digital signature for the whole stream and hashes for each block of "*c*" packets. The receiver required a buffer of size "*c*". The receiver first received the signature of the 20-byte hash of the first block and the hash itself. After verifying the signature, the first hash should be verified. He then started receiving the first block and calculated the hash for this block. If it matched the verified hash, it could then play the block. Otherwise the whole stream was rejected. The first block carried the hash for the second block, and so on.

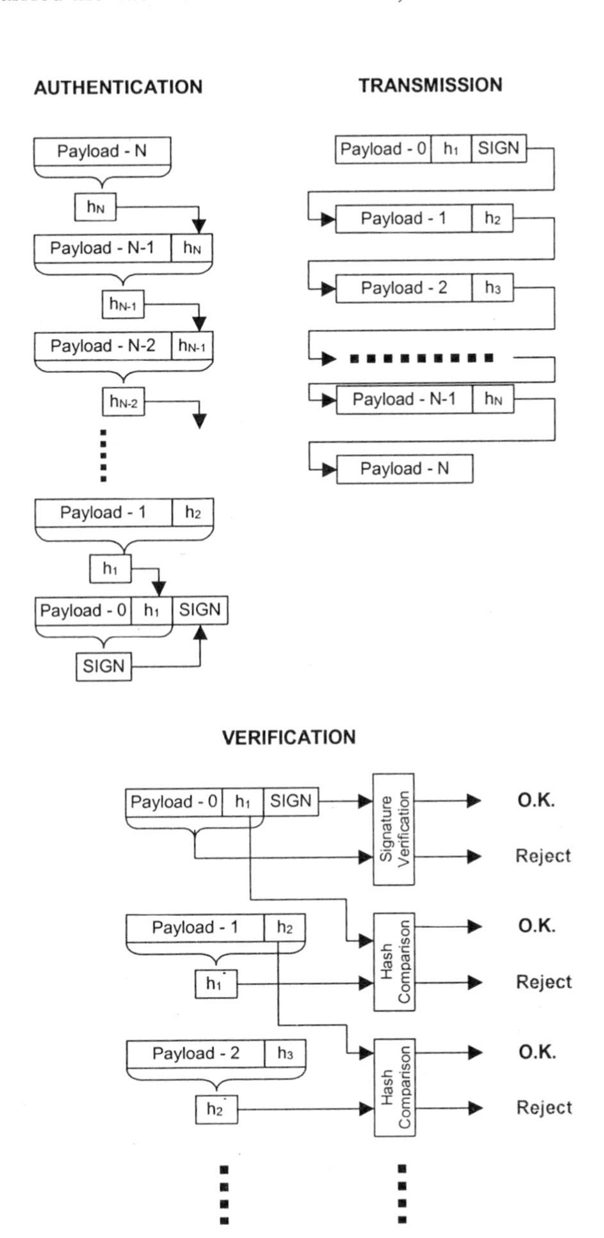

Figure 10-6. Illustration of the hash chain of Gennaro and Rohatgi; h_i – hash of packet P_i, h^*_i – hash calculated for the received packet P^*_i.

At the sender, the whole stream had to be known in advance, as the hash for the block "i+1" was appended to the block "i". Reducing the block size "c" to a single packet meant no need for a buffer at the receiver side. The schematic diagrams explaining the authentication procedure, transmission and verification are given in Fig. 10-6

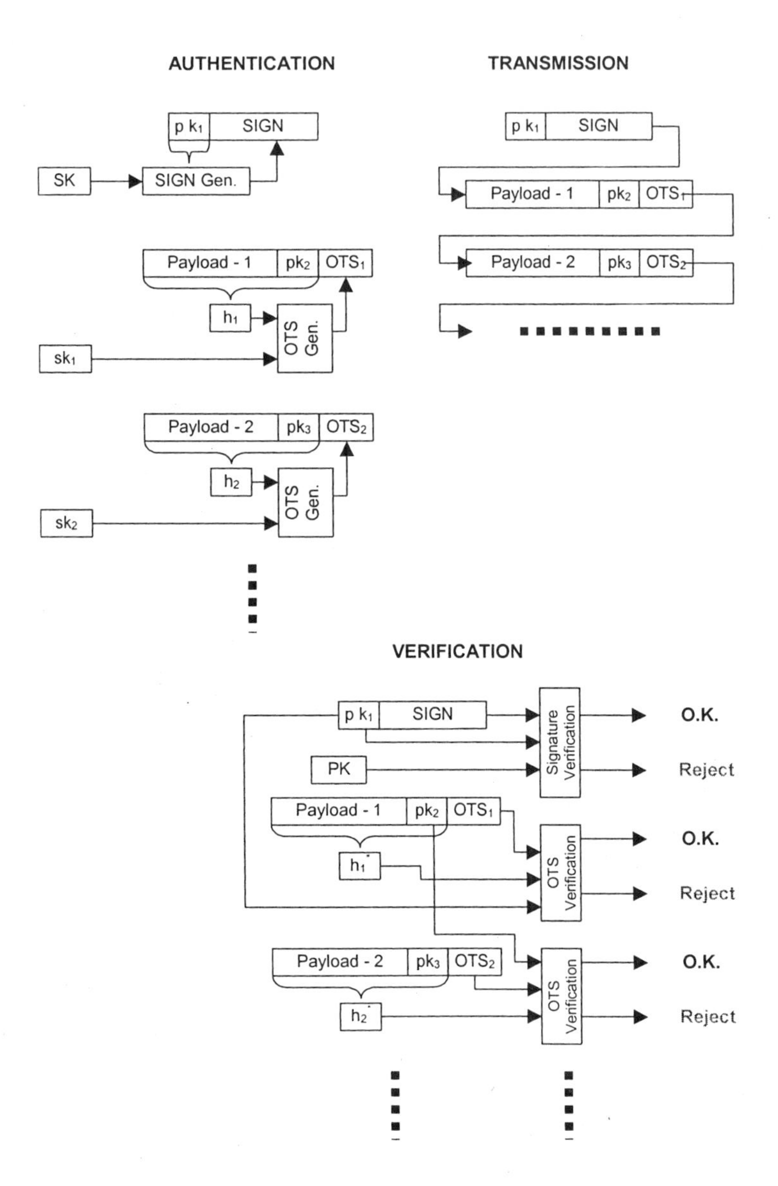

Figure 10-7. Illustration of the one-time signature chain of Gennaro and Rohatgi; *SK* – secret key, sk_i – one-time secret keys, *PK* – public key, pk_i – one-time public keys.

The main advantages of the scheme were very low computational and communicational overheads. However, the scheme was very vulnerable to packet losses, and even a single packet loss would result in a break in the verification chain, and rejection of all successive packets. Moreover, the scheme was suited for the off-line applications only.

For the case when the entire data stream is not known in advance, Genaro and Rohatgi propose a scheme [5] based on One–Time Signature (OTS) [12]. These signatures are much faster to compute and verify but can only be used for a prefixed number of messages, usually one. The sender starts with sending a signed public key for an OTS scheme. Than the sender sends the first block with the appended OTS computed based on its hash to be verified using the one-time public key sent previously. The first block contains also a new one-time public key to be used to verify the signatures on the second block, and so on. The scheme is illustrated in Fig.10-7.

Later in [5], Gennaro, Rohatgi introduce a simplification to the scheme using a family of Target Collision Resistant (TCR) hash functions. As a result, the size of communication overhead can be significantly reduced without compromising the security. Also the computational overhead is much lower. Unfortunately, its vulnerability to packet losses is maintained.

3.2 Augmented Hash Chains

Several modifications to the Gennaro and Rohatgi schemes were proposed in literature, and well generalized by Miner and Staddon in [11]. Miner and Staddon propose there to introduce additional connections in the authentication chain in order to achieve immunity against lost packets. As the result, the verification chain can tolerate even burst losses of up to the predefined number of packets. However, once the chain is broken due to the longer than assumed burst, the verification cannot be continued. Another disadvantage of the scheme is the fact that it is suited for the off-line authentication only, as the sender needs to know all packets in the stream to calculate the desired sequence of hashes.

4. STREAM AUTHENTICATION FOR SYNCHRONOUS SYSTEMS

In [9], Perrig et. al. have introduced a group of five schemes for stream data authentication. All of those schemes begin with an initial synchronization of the sender's and receivers' clocks, and the receivers need to record an upper bound on their clock differences with the sender's clock. The underlying assumption of all these schemes is that during the transmission of the stream, the local internal clocks of the sender and receiver do not drift too much. The five schemes presented in [9] are all related to each other, and they are introduced in a way that with the increased complexity, better properties are obtained.

The basic authentication scheme in [9] involves the following algorithm. To send the message M_i, $i = 2, 3, \ldots$, the sender assumes that the receiver has an authenticated packet $P_{i-1} = \langle D_{i-1}, MAC(K'_{i-1}, D_{i-1})\rangle$, where $D_{i-1} = \langle M_{i-1}, F(K_i), K_{i-2}\rangle$, and M_{i-1} is the message contained by packet P_{i-1}, $F(K_i)$ commits the key K_i without revealing it, $K'_i = F'(K_i)$ is the secret key used to compute the message authenticating code (*MAC*) of the next packet. The functions F and F' are two different pseudorandom functions that are impossible to be inverted by an adversary.

The sender picks a fresh random key K_{i+1} and constructs the packet $P_i = \langle D_i, MAC(K'_i, D_i)\rangle$, where $D_i = \langle M_i, F(K_{i+1}), K_{i-1}\rangle$. The receiver can only verify packet P_i after receiving packet P_{i+1}, as it does not know K_i and cannot reconstruct K'_i. By computing $F(K_i)$ the receiver can verify the key K_i since $F(K_i)$ has been received in P_{i-1} and then compute $K'_i = F'(K_i)$. It can then verify packet P_i by checking its *MAC*. Authentication of P_i authenticates also the commitment $F(K_{i+1})$ necessary to authenticate P_{i+1} after P_{i+2} is received.

The first received packet ($i = 1$) must be authenticated with a regular digital signature scheme, like *RSA* [8].

This simple verification scheme is, of course, very susceptible to packet loss and can be attacked if the attacker gets packet P_{i+1} before the receiver gets packet P_i. Such a scenario would allow the attacker to know the secret key K_i used to compute the *MAC* of P_i and change the message and the commitment in b P_i. As a result, all subsequent traffic could be forged.

The basic scheme is then improved in [9] through the introduction by the sender of the pre-computed key chain $\{K_i\}$, such that $K_0 = F^n(K_n)$, and $K_i = F^{n-i}(K_n)$, where υ consecutive applications of the pseudorandom function F are denoted as $F^{\upsilon}(x) = F^{\upsilon-1}[F(x)]$, and $F^0(x) = x$. That way, the attacker knowing K_i cannot compute K_j for $j > i$. On the other hand, the receiver can compute all K_l from K_i if $l < i$, since $K_l = F^{i-l}(K_i)$. Therefore, if the receiver received packet P_i, any subsequently received packet will allow it to compute K_i and $K'_i = F'(K_i)$ and verify authenticity of P_i. Hence, this modified scheme can tolerate an arbitrary number of packet losses (packets rejected due to corruption).

Three other modifications to the basic authentication scheme proposed in [9] deal with allowing faster packet transfer rates (not necessaryly limited by the slowest of the receivers), dynamic packet rates, and introduction of the mechanism to facilitate accommodation of a broad range of receivers with different timing requirements. The most advanced scheme, referred to as the Time Efficient Stream Loss-tolerant Authentication (TESLA) scheme, is almost ideally suited to operational conditions in mobile networks, where the initial synchronization of the sender with all receivers has to be maintained for a proper functioning of the whole system. The most prominent examples of synchronization requirements are the necessity for a perfect synchronization of the transmitter and the receivers when the time division multiple access (TDMA) is used (e.g. GSM/GPRS networks) or in the case of orthogonal frequency division multiplexing (OFDM) as in the wireless networks compliant with IEEE802.11a/g standards.

The advantages of the TESLA scheme are as follows [9]:

- Low computational overhead.
- Low per-packet communication overhead.
- Tolerance to arbitrary packet loss.
- No acknowledgements required.

The main disadvantages of the TESLA scheme are [9]:

- No guarantee of non-repudiability.
- Requirement for an initial synchronization of the sender with receivers.

5. GRAPH BASED AUTHENTICATION SCHEMES

The two main authentication schemes, which use a graph-basedapproach to construct the augmented chain of hashes and achieve non-repudiation for on-line transmission through periodic signature packets are the signature scheme of Perrig et.al. [9] and the authentication scheme introduced by Golle and Modadugu [10]. As both schemes are very similar, we limit our considerations here to the one proposed in [10]. Because of the graph representation, their performance in the presence of packet losses can be assessed using the Markov chain approach. In particular, the Markov chain analysis is very convenient in cases where the virtual transport channel can be modeled using the Gilbert-Elliot two-state [14] or a more advanced hidden Markov model.

In this section, we will first describe the basic Golle and Modadugu augmented hash chain. Then, using the Gilbert-Elliot model we will show that in realistic scenarios, one cannot neglect the possibility that a burst of lost or corrupted packets include a signature packet. This will be followed by a description of a modification proposed in [12], which can overcome the problem of losing a signature packet for the schemes introduced in [9] and [10].

5.1 Golle and Modadugu authentication scheme

The technique proposed by Golle and Modadugu in [10] follows from the fact that if a collision-resistant hash of packet P_1 is appended to packet P_2 before signing P_2, then the signature on P_2 guarantees the authenticity of both P_2 and P_1. In general, a hash h_1 of P_1 is appended to P_2 before calculating a hash h_2 for P_2, and h_2 is appended to P_3 before calculating h_3 for P_3, and so on. The final packet in a sequence, P_n is then signed after appending h_{n-1} to it. If then the sequence of packets P_1, ..., P_n is received without tampering or losses, all packets in that sequence can be authenticated. The process of creating such a simple authenticating graph is presented in Fig. 10-8.

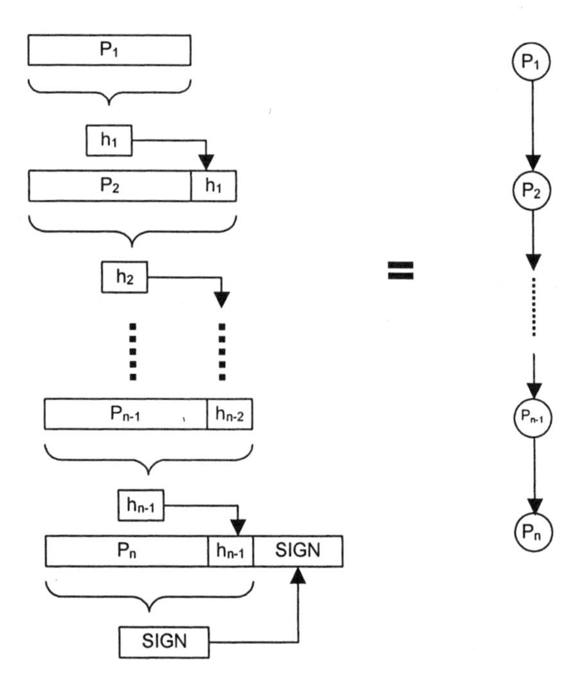

Figure 10-8. Basic authentication chain of Gole and Modadugu and the corresponding authentication graph [12].

The simple chain presented in Fig. 10-8 can be modified to include supplementary connections to prevent it from being broken in a case of packet losses. Depending on the type of construction of those additional connections, immunity from burst losses can be achieved. As an example, in Fig. 10-10 (a), the chain immune to a loss of two consecutive packets is presented. Several different constructions are presented in [10] and analyzed.

The main advantage of the scheme is the fact that by splitting the stream into smaller sequences of packets and reversing the order of the authentication chain, Golle and Modadugu achieved a scheme suited for on-line applications where a break in the verification chain means rejection of usually just one sequence of packets (maximum two sequences if the break includes a signed packet). Other advantages of the technique are immunity to bursts of up to a given length, low communication overhead, and low computational overhead [10]. The drawbacks of the scheme are the delayed verification and susceptibility to a loss of signed packets.

5.2 Gilbert-Elliot channel model

In his fundamental paper [14], Gilbert introduced a two state Markov chain to model a transmission channel with burst errors. The model has been later refined by Elliott in [15], and is generally known in telecommunications literature as the Gilbert-Elliott channel. The Gilbert-Elliott model has been introduced to analyze a physical channel. However, we can use the same approach to perform analysis of the virtual transport channel.

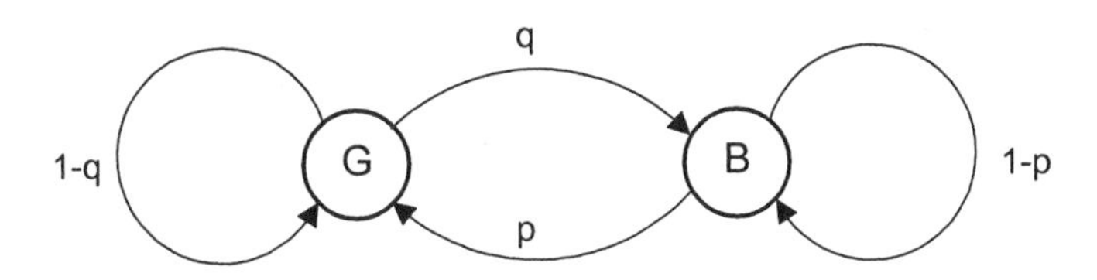

Figure 10-9. Gilbert-Elliot channel model.

Gilbert in [14] has shown that a Markov chain with two states can be used to generate bursts. The model is shown in Fig. 10-9, where the states *G* and *B* denote the "good" channel state and the "bad" channel state, respectively. In the "good" state the probability of packet loss approaches zero, while in the "bad" state it can take any arbitrary value greater than zero. In the original model [14], that probability was set to 0.5, as the author dealt with bursts at the bit level, i.e. in the physical channel, where bursts contain good bits interspersed with the errors. In [12], to facilitate considerations, the authors assumed that at the packet level that probability is very high, approaching 1. The model is described by the probability transition matrix $\mathbf{P}_T$ given by:

$$\mathbf{P}_T = \begin{bmatrix} 1-q & q \\ p & 1-p \end{bmatrix} \tag{10.1}$$

and the corresponding graph is presented in Fig. 10-9.

For this Markov chain, the stationary probability vector $\mathbf{P}_S = [p_1, p_2]$ can be calculated using the formula:

$$\mathbf{P}_S = \mathbf{P}_S \mathbf{P}_T \tag{10.2}$$

and the normalization condition:

$$p_1 + p_2 = 1 \tag{10.3}$$

From (10.2) and (10.3), we get a set of two simultaneous equations:

$$\begin{cases} p_1 = p_1(1-q) + p_2 p \\ 1 = p_1 + p_2 \end{cases} \tag{10.4}$$

Solving it for p_1 and p_2 gives:

$$p_1 = \frac{p}{p+q} \text{ and } p_2 = \frac{q}{p+q} \tag{10.5}$$

The stationary probabilities p_1 and p_2 are the probabilities that at any discrete time instant the channel is in the "good" state or the "bad" state respectively. For the transmitted packets, this translates on the probabilities that the packet is either received correctly or lost. In addition, the exact values of the probabilities p_{12} and p_{21} fully determine the probabilities of bursts occurrences and their lengths.

Let now denote by $P_g\{M\}$ the probability that out of M signed packets in the stream all packets have been transmitted in "good" state, and that subsequently no such a packet has been lost. Following the above analysis of the Gilbert-Elliott model, we can write that:

$$P_g\{M\} = p_1^M \tag{10.6}$$

Hence, the probability $P_b\{1\}$ that at least one signed packet is transmitted in the "bad" state and subsequently lost is given by:

$$P_b\{1\} = 1 - P_g\{M\} = 1 - p_1^M \tag{10.7}$$

An example analysis presented in [12] shows that in a realistic scenario, even for quite a modest stream of 50 sequences of 50 packets (for packets of 1000 bytes this corresponds to about 2.5MB), the probability of losing at least one signed packet is very high, and with the increased number of sequences it approaches 1.

5.3 Modified Golle and Modadugu scheme

In [12], a simple method to overcome the problem of losing a signed packet is proposed without introducing any additional overheads or delays in packet verification, when no signature is lost. An extra delay in verification time will be only introduced in the case of a burst containing the signed packet. However, in the original scheme described in [10], this situation would result in the whole considered sequence of packets being rejected. Of course, the method introduces additional buffer requirements at the receiver.

The method proposed in [12] is based on extending the authentication chain of [10] beyond the packet containing the signature, i.e. the packet, on which the authentication graph should end for the given sequence. In other words, the signed packet is considered in a similar way to any other packet, with the exception that no hash is generated for it.

Suppose a stream is divided into sequences S_1, S_2, ..., where each S_k consists of n packets. Let

$$S_k = \{P_{(k-1)n+1}, P_{(k-1)n+2}, \ldots, P_{kn}\} \tag{8}$$

For each sequence Golle and Modadugu [10] proposed an optimal authentication chain to provide resistance for bursty loss of packets with the assumption that the signed packet P_{kn} is not lost. Their authentication chain is modified by introducing extra edges between S_k and S_{k+1} and removing the assumption of no signed packet lost.

The modified authentication chain is defined in [12] as follows. Let a be a positive integer. The hash of a packet P_i is appended to two packets P_{i+1} and P_{i+a} for all i, $(k-1)n < i < kn$, and for all $k \geq 1$. This authentication chain sustains bursts of $a - 1$, packets, even including the packet with signature. If the signature packet is lost, the authentication has a delay of $2n$.

An example application of the proposed solution together with the authentication chain of [10] is given in Fig. 10-10 [12]. The diagram of Fig. 10-10(a) shows the original authentication chain, while the diagram of Fig. 10-10(b) presents what happens around the signed packet P_n. As the authors of [10] point out, their authentication chain (Fig. 10-10(a)) is immune to bursts of 2 packets, but the signed packet P_n must be delivered correctly. In the modified case, if the burst of 2 packets contains the signed packet, e.g. packets P_{n-1} and P_n are lost, the authentication chain is not broken, and verification can be performed after receiving the next signed packet, i.e. P_{2n}. Moreover, if the bursts are no longer than the length for which the original chain has been designed, any received signature can be used to verify all previously received packets, no matter how many signed packets have been lost.

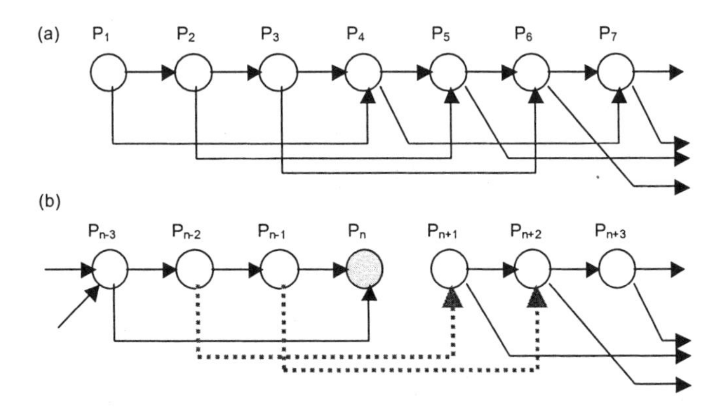

Figure 10-10. Explanation of the modification method proposed in [12]; (a) the original authentication chain of order 3 [10], (b) continuation of the chain beyond the signed packet P_n.

There are no additional computational overheads involved, and there is almost no additional communication overhead involved in the scheme proposed in [12] compared to the original scheme described in [10] (the hashes from the previous sequence packets are now attached to the current sequence packets, as in Fig. 10-10(b)). All chain constructions proposed in [10] can be used together with the proposed method. The same technique can be extended to the signature scheme proposed in [9].

6. COMPARISON OF STREAM AUTHENTICATION TECHNIQUES

In the chapter, we presented some of the most prominent methods proposed in literature for authenticating digital streams. All of the schemes have their advantages and disadvantages. They are compared in Table 10-1, where we look into the following features: on-line/off-line operation, requirement for synchronization, processing overhead, communication overhead, non-repudiation, delay in authentication, immunity to packet loss, and security. There are, of course other features, like for example requirement for buffering of packets at both sender and receiver sides, which need to be taken into account when deciding about the choice of a scheme.

Table 10-1. Comparison of stream authentication schemes.

Scheme	Operation	Synch. required	Processing overhead	Comm. overhead	Non-repudiation	Delay	Immunity to losses	Security level
Star chaining [6]	On	No	H	H	Yes	S	Yes	H
Tree authentication [6]	On	No	H	H	Yes	S	Yes	H
Hash chain [5]	Off	No	S	S	Yes	S	No	M
One-time sign. [5]	On	No	M	M	Yes	S	No	M
TESLA [9]	On	Yes	S/M	S	No	S/M	Yes	M
Hash chain [10]	On	No	S	S/M	Yes	M	Yes	H
Signature scheme [9]	On	No	S/M	S/M	Yes	M	Yes	H

Note:

On = 'On-line', Off = 'Off-line', H = High, M = Medium, S = Small

REFERENCES

1. W. Stalings: *Data and Computer Communications*, 6th ed., Prentice Hall, Upper Saddle River, 2000.
2. J.Lu: "Signal Processing for Internet Video Streaming," Proc. of SPIE, Image and Video Communications and Processing, January 2000.
3. G. Pipkin: http://www.itc.virginia.edu/atg/techtalks/powerpoint/video/sld001.htm "Video on the Web," (accessed on 23/11/01).
4. Utah Education Network - http://www.uen.org/technical/html/streamingfaq.html (accessed on 23/11/01)
5. R.Gennaro, P.Rohatgi: "How to Sign Digital Streams," *Cypto '97*, Springer-Verlag, 1998, pp. 180-197.
6. C. K. Wong, S. S. Lam: "Digital signatures for Flows and Multicasts," *IEEE/ACM Transactions on Networking*, Vol.7. No.4, Aug.1999, pp. 502-513.

7. M.J. Moyer, J.R.Rao and P.Rohatgi: "A Survey of Security Issues in Multicast Communications," *IEEE Network*, Nov/Dec.1999, pp. 12-23.
8. D. R.Stinson: "*Cryptography Theory and Practice*," CRC Press.
9. A.Perrig, R.Canetti, J.D. Tygar, D. Song: "Efficient Authentication and Signing of Multicast Streams over Lossy Channels," *Proc. of IEEE Security and Privacy Symposium*, May 2000, pp. 56-73.
10. P.Golle, N.Modadugu: "Authenticating Streamed Data in the Presence of Random Packet Loss", Proc. of Network and Distributed System Security Symposium, 8-9 February 2001.
11. S.Miner, J.Staddon: "Graph-Based Authentication of Digital Streams," http://www-cse.ucsd.edu/users/sminer/abstracts/GraphAuth.html (accessed on: 11/01/02).
12. B.J. Wysocki, Y. Wang, and R. Safavi-Naini: "On a Method to Authenticate and Verify Digital Streams," *Journal of Telecommunications and Information technology*, vol. , No.2, pp. pp.45-52, 2002.
13. W. Stalings: "*Cryptography and Network Security: Principles and Practice*," 2nd ed., Prentice Hall, Upper Saddle River, 1998.
14. E. N. Gilbert, "Capacity of Burst-Noise Channel", The Bell System Technical Journal, Sept. 1960, pp.1253-1265.
15. E. O. Elliott, "Estimates of Error Rates for Codes on Burst-Noise Channels", The Bell System Technical Journal, Sept. 1963, pp.1977-1997.

Chapter 11

FEATURES OF PARALLEL TCP WITH EMPHASIS ON CONGESTION AVOIDANCE IN HETEROGENEOUS NETWORKS

Qiang Fu and Jadwiga Indulska
School of Information Technology and Electrical Engineering, The University of Queensland, Brisbane, QLD 4072 Australia, {qiangfu, jaga}@itee.uq.edu.au

Abstract: In this chapter we describe and model Parallel TCP, an extension of the TCP protocol, designed for heterogeneous networks with wired and wireless (possibly multihop) links. Parallel TCP splits a standard TCP connection into a number of parallel virtual connections. An analytical and a simulation model are presented and used to evaluate the performance of Parallel TCP. It is shown that based on these models Parallel TCP could improve TCP performance in heterogeneous networks by either dynamically adjusting the number of virtual connections or adjusting the Congestion Avoidance (CA) algorithm in a static number of connections. The latter is a preferable solution as dynamically changing the number of connections creates a large connection management problem.

Key words: Congestion avoidance, heterogeneous networks, protocol modeling, parallel connections, TCP, wireless networks

1. INTRODUCTION

The rapid growth and worldwide expansion of the Internet has meant that the future Internet will have a large number of very high-bandwidth links and increased diversity in network access technologies. The increasing popularity of wireless networks indicates that error-prone wireless links will piay an important role in future networking. These wireless links will significantly increase random packet loss and thus degrade TCP performance [1], because the packet loss on wireless links will trigger unnecessary congestion control responses. Recent research has shown that, even for wired networks, the throughput of high performance applications using TCP is persistently and significantly lower than the bandwidth available on the network [2]. One source of the poor TCP throughput that has been identified is that the actual packet loss rate is much greater than what would be reasonably expected [3], [4]. On the other hand, the basic Additive Increase Multiplicative Decrease (AIMD) algorithm [5] is currently not working well on the networks with a high bandwidth-delay product (BDP). With large windows, additive/linear increase by one segment per Round-Trip Time (RTT) could be too slow, while multiplicative decrease per loss event could be too drastic.

A considerable amount of research effort has been directed at reducing unnecessary congestion control responses in TCP to random packet loss and delay [6]. The existing TCP enhancements focus on distinguishing random packet loss from congestion-triggered packet loss. However, it is not always possible to provide such a distinction due to the asymmetry of the round trip path, unstable cross traffic and network dynamics, insufficient information to identify corruption errors, etc. [7]. In Mobile IP [8], the indirect routing on the forward path and the direct routing on the reverse path would degrade the performance of RTT measurement based TCP schemes such as TCP Vegas [9], ST-PD TCP [10], etc. [11]. Some of these enhancements are fine-tuned for wireless links, like proxy-based solutions which hide the wireless part from the sender and thus provide local reliability. However, these solutions have limitations. For instance, the forward and reverse paths of a TCP connection must pass through the same proxy, and it is usually assumed that there is only one wireless hop between the proxy and the mobile host and, accordingly, aggressive local recovery policies are adopted over the single wireless link. However, if there are multiple wireless hops, these solutions may not be effective. Moreover, for some solutions encryption at the TCP level is impossible and buffer space at the proxy may run out due to long disconnections.

The recent focus on the performance of high-BDP networks has resulted in the development of TCP variants such as Scalable TCP [12], HighSpeed TCP [13], XCP [14] and FAST TCP [15]. XCP, which is based on the Explicit Congestion Notification proposal [16], is a redesign of the congestion control architecture which does not care about backward compatibility or deployment. Compared to their respective prototypes – TCP Reno and Vegas – the other three variants are more aggressive when increasing window size and/or less drastic when decreasing it, to suit high-BDP networks. They can achieve promising throughput on high-BDP networks, however they could face problems when used in heterogenous networks.

Parallel TCP is a TCP enhancement that splits a standard TCP connection into a number of parallel virtual connections. In other words, several virtual

TCP connections are created instead of a standard TCP connection for an application-layer connection. Parallel TCP can use a single-socket operation for all virtual connections and central management of individual congestion windows for virtual connections, in order to improve TCP performance. Parallel TCP (similar to Scalable TCP, HighSpeed TCP and FAST TCP) can cater for high-BDP networks. Moreover, some features of Parallel TCP, such as self-curing, make it less sensitive to random and transient packet losses. However, Parallel TCP also faces some challenges. As the number of virtual connections increases, Parallel TCP connections may lead to congestion and they may steal bandwidth from standard TCP connections. An intuitive response to these situations is to dynamically change the number of connections [17], or make parallel TCP connections behave like a single standard connection as far as congestion control is concerned [18], [19]. In addition to multiple connections and sluggish congestion control, Congestion Avoidance (CA) [20] is another important factor which can make parallel TCP connections aggressively grasp bandwidth. For example, if there are n virtual connections and each of them adopts the standard CA algorithm, then the aggregate TCP window increases by n segments per round trip, instead of 1 segment per round trip, which is the case for a standard TCP connection. Therefore, it is worth investigating the effects of CA on Parallel TCP. Dynamically adjusting the CA algorithm could be an alternative solution to, or could possibly complement, the previously outlined approaches.

This chapter presents the motivation for Parallel TCP and evaluates its performance. It also discusses how changes in CA methods affect the performance of Parallel TCP. To illustrate the effects of CA, a simple performance model, which takes CA into account, is presented for Parallel TCP and the impact of CA on its performance is explored.

The chapter is organized as follows. Section 2 describes related work. Section 3 presents an overview of Parallel TCP. In Section 4, the role of CA is examined using the bandwidth estimation model developed for Parallel TCP. In Section 5, the impact of Congestion Avoidance is presented based on simulation and Section 6 concludes the chapter.

2. RELATED WORK

There have been efforts towards improving TCP performance by opening parallel TCP connections on a set of sockets. Some of these TCP extensions are implemented at the application layer, the others at the transport layer. [21], [22] and [17] describe solutions which increase FTP throughput by establishing multiple FTP connections to overcome the limitation on the

TCP window size in satellite-based environments. In [2], an enhanced FTP protocol called GridFTP is designed for the grid environment. [23] proposes an end-to-end transport layer approach, called pTCP, to perform bandwidth aggregation on multi-homed mobile hosts. These hosts have subscriptions and access to more than one wireless network at a given time. [18] explores the use of multiple independent concurrent TCP flows, and shows that an ensemble of concurrent TCP connections can effectively share bandwidth and obtain consistent performance without adversely affecting other network flows. [24] introduces PSockets, a library developed to deliver data over multiple TCP connections with dramatically increased performance on a poorly tuned host compared to the performance of a single TCP stream. Based on [25], [26] develops a theoretical model of parallel TCP connections. The model is basically a sum of the common bandwidth estimation for each single TCP connection developed by [25]. The model has four factors, namely, the number of connections, the Maximum Segment Size (MSS), the Round Trip Time (RTT) and the packet loss rate. [27] discusses the issues regarding fairness, effectiveness and efficiency of parallel TCP connections and shows their positive effects.

All of the approaches described above are based on multiple-socket operations. Some of them adopt a method in which the individual connections are independent to each other and thus adjust their congestion window separately. This lack of cooperation may have a negative impact on performance. Other approaches do adopt integrated congestion control across parallel connections, such as [18] and [19], where an integrated congestion control is as aggressive as a single standard TCP connection. However, there is not enough evidence to suggest that the aggregated window should be halved when a packet loss occurs on any of the parallel connections [28]. In this chapter an implementation of the single-socket operation is discussed. Central management of individual congestion windows is suggested to improve TCP performance. This is done by allocating the same window to all the virtual connections that can achieve improved throughput performance in most cases.

SCTP was originally designed to provide a general-purpose transport protocol for message-oriented applications, like the transportation of signaling data [29]. Its multi-streaming function allows data to be partitioned into multiple streams, each with independently sequenced delivery. Message losses in any one stream only affect delivery within that stream, not delivery in other streams. However, the flow control and congestion control in SCTP remains similar to the standard TCP.

Scalable TCP [12] is a simple modification to the standard TCP congestion control algorithm. It uses Multiplicative Increase Multiplicative Decrease (MIMD) with constant parameters. After a loss event, its window opens exponentially instead of linearly. For instance, when there is no congestion detected, for each acknowledgment received in a RTT the window increases by 0.01 segments (1/*cwnd* for standard TCP). On the first detection of congestion in a given RTT, the window is cut by 0.125*cwnd* (0.5*cwnd* for standard TCP). Similar to Scalable TCP, HighSpeed TCP [13] involves a subtle change in the TCP response function. It uses the AIMD algorithm with the AI and MD parameters that are increasing functions of the current window size where AI=*a*(*cwnd*)/*cwnd* and MD=*b*(*cwnd*)**cwnd*. Both Scalable TCP and HighSpeed TCP use the parameter *Low_Window*. They use the same response function as standard TCP when the current congestion window is at most *Low_Window*, and use the modified (aggressive) response function when the current congestion window is greater than *Low_Window*. At a window size larger than *Low_Window*, both behave like an aggregate of *N* TCP connections. FAST TCP [15] is based on TCP Vegas, but more suitable for high speeds and large-BDP flows. FAST TCP roughly emulates the congestion control response of *N* parallel Vegas TCP connections, where *N* increases as a function of the current throughput (*cwnd*/*RTT*).

All the three TCP variants (Scalable, HighSpeed and FAST) have the robustness/aggressiveness of parallel connections without introducing multiple connections. However, some benefits of multiple connections are not available, for example, self-curing (when a connection is disrupted the other connections can use the released bandwidth from the disrupted connection). Another example is the property of independently sequenced delivery, also used by SCTP, which at the receiver limits the affect of out-of-order and lost segments to the delivery within the involved connections. On the other hand, maintaining a large window may not be a good idea: some operating systems do not provide good support for large windows; large windows may not be supported through some network components such as firewalls; and large windows increase the possibility of multiple packet losses in a single window. Furthermore, the Internet can be thought of as a black box system where it is not an easy task optimizing parameters to achieve the desired performance. The features of these TCP variants make them promising for high-BDP networks but less suitable for heterogeneous wired-wireless links, including multihop wireless links.

In [17], a dynamic change of the number of connections (based on the congestion control method of TCP Vegas) is proposed to avoid congestion while maximizing throughput. Dynamically changing the number of connections could increase the complexity of connection management, especially for single-socket operations. Furthermore, this method does not take random packet loss into account. In systems with a high rate of random packet loss, maintaining a certain number of connections is usually preferred.

This chapter shows that dynamically adjusting the CA algorithm could be used as an alternative to the dynamic change of the number of connections. To show the impact of CA on parallel TCP connections, we introduce a simple model that not only takes the number of connections, the MSS, the RTT and the packet loss rate into account but also the Congestion Avoidance, the one-connection window (W_1), the multiple-connection window (W_n) and the maximum available window (W_m). A generic description of the model is presented in [30] and [31]. To comply with the consumption of constant RTT that is also used by [25] and thus [26], the model has a particular analysis of the situation when $W_n>=W_m$. The performance analysis focuses on the effects of random packet loss.

3. PARALLEL TCP

The design of Parallel TCP is motivated by the following considerations:

1. Initial Window (IW), the initial value of the Congestion Window (*cwnd*) may not be necessarily set to less than two segments. A larger IW can provide quicker increases in the size of *cwnd* during the Slow Start stage. Some proposals have demonstrated the advantage of a larger IW [32], [33]. Parallel TCP increases IW to n segments, i.e. the number of virtual connections if all the virtual connections start transmitting at the same time. However, it is not necessary to start all the virtual connections at exactly the same time, as congestion could be created during Slow Start.
2. When a packet is lost, delayed or reordered, only the involved TCP connection is disrupted while the others are not affected.
3. At the receiver, the independently sequenced delivery (also used by SCTP) limits the affect of out-of-order and lost segments to the delivery within the involved connections. This feature is of particular significance for high-BDP networks.
4. When multiple packets are lost from one window of packets, traditional TCP may experience poor performance. However, Parallel TCP can decrease the likelihood of multiple packet losses since the packet losses are dispersed across a number of virtual connections.
5. By using Parallel TCP, a large aggregate window size at the receiver can be used to overcome unnecessary limitations on the window size imposed by TCP flow control or other factors. The limitation restricts the throughput on networks such as satellite networks that have a high bandwidth-delay product.

6. Parallel TCP introduces the feature of self-curing. When a connection is disrupted due to unnecessary congestion control responses and therefore releases its bandwidth, the other virtual connections can utilise the released bandwidth.
7. The diversity of networks creates a need for a TCP solution robust across a wide range of environments rather than fine-tuned for one particular environment or traffic type at the expense of another.

Parallel TCP can be built on the existing TCP schemes, which comply with TCP end-to-end semantics, to enhance their performance. Parallel TCP provides a connection manager to deal with issues such as packet allocation to virtual connections, buffer and window management, etc. The virtual connections can change CA schemes and can be started and closed dynamically during the application life-time to reflect the network dynamics. At the sender, the virtual connections could ask for packets when their individual buffers are not full. If each virtual connection is provided a separate socket (multiple-socket operation), then a unique sequence number is required for each packet to identify its original location in the packet stream.

It is possible that all the virtual connections share a single socket (single-socket operation). For instance, let n be the number of connections, connection 1 transfers packets whose indices are 1, $1+n$, $1+2n$, $1+3n$, and so on while connection 2 transfers packets whose indices are 2, $2+n$, $2+2n$, $2+3n$, etc., and thus the sequence number for each packet is from the same series of continuous sequence numbers. As all the virtual connections may not be able to achieve the same throughput, the slowest connection may significantly slow down the whole transfer process. This problem can be addressed by allocating the same window to all the virtual connections. The window equals the aggregate window size divided by the number of connections. If there is still a significant throughput difference between connections, increasing the window size of the slow connections while decreasing the window size of the fast ones can be a remedy. On the other hand, frequently changing the number of connections increases the operation complexity at both sender and receiver sides as packets belonging to a given connection may be allocated to another connection after a change of the number of connections. This situation complicates the single-socket operation. Therefore adjusting the CA algorithm can be an alternative to dynamically changing the number of connections. In the following sections we will discuss this in detail.

4. CA IN THE BANDWIDTH ESTIMATION MODEL

A simplified model for Parallel TCP is presented in this section. The model is intended to demonstrate the relationships between the throughput and the CA, the number of connections, the packet loss rate, the one-connection window (W_1), the multiple-connection window (W_n), and the maximum available window (W_m) which is defined by the bandwidth-delay product. The model assumes that TCP is running over a lossy path and the congestion window (W_1 or W_n) is equal to or less than W_m, and thus a constant round trip time (RTT) can be achieved. The model assumptions are as follows:

1. TCP is fully controlled by CA [20]
2. Constant Round Trip Time (RTT)
3. Single packet loss in each saw cycle
4. No errors on the reverse path
5. Fixed packet size (MSS)
6. Random packet loss at a constant probability p
7. Fast Recovery is ignored.

It should be noted that W_n=W_1 when there is only one virtual connection, W_m equals the bandwidth-delay product and the theoretical W_n can be larger than W_m. The following subsections discuss both cases of $W_n < W_m$ and $W_n >= W_m$.

4.1 Case of $W_n < W_m$

With the above model assumptions, the TCP congestion window changes according to a perfectly periodic sawtooth shape [25]. Fig. 11-1 shows the TCP congestion window pattern based on one virtual connection. W_1 is the maximum value of the window, that is, W_1 packets. According to the standard Congestion Avoidance algorithm, the minimum window must be half the size of the maximum window, that is, $W_1/2$ packets. If the receiver is acknowledging every packet, then the window increases by one packet per round trip, so each cycle must be $W_1/2$ round trips, that is, $W_1/2*RTT$ seconds. The total data delivered in each cycle must be $1/p$ (assumption 6) packets, as the TCP window is cut down to half when a packet is lost and a packet loss happens after $1/p$ consecutive packets are delivered. Therefore, we have:

$$(W_1/2)^2 + (1/2)(W_1/2)^2 = 1/p \qquad (11.1)$$

Solving for W_1 we get:

$$W_1 = \sqrt{8/3p} \tag{11.2}$$

If we substitute W_1 in the following bandwidth equation, then we get the estimated throughput based on one connection:

$$BW_1 = \frac{(1/p) * MSS}{(W_1/2) * RTT} = \sqrt{(3/2)/p} \times \frac{MSS}{RTT} \tag{11.3}$$

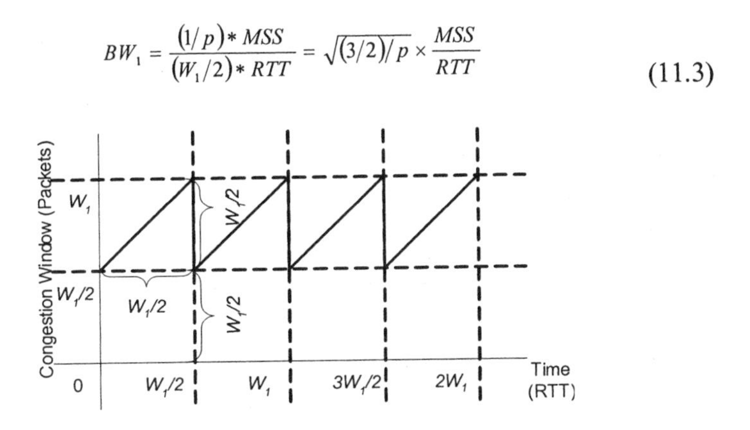

Figure 11-1. Sawtooth TCP window evolution for periodic loss

Based on a similar idea, we now introduce n virtual connections (Fig. 11-2). A further assumption is that packet losses happen evenly on these connections. Suppose the aggregate window for these n connections is W_n, and assume that W_n is always less than the bandwidth-delay product, the maximum available window (W_m). Then the window for each connection is W_n/n. When a packet loss happens, only the involved window is reduced by $W_n/2n$. The other connections stay free from the interruption. Therefore, the minimum window is $(2n\text{-}1)W_n/2n$. For each virtual connection, the window increases by 1 packet (segment) per round trip and thus it takes $W_n/2n$ round trips for the window to increase by $W_n/2n$ packets. Therefore for n connections, the aggregate window opens by n packets per round trip and it takes $W_n/2n^2$ round trips for the window to increase by $W_n/2n$ packets. That is, each cycle is $W_n/2n^2 * RTT$ seconds.

The above analysis is based on the standard CA algorithm, which is adopted by each virtual connection. That is, the congestion window for each connection opens by 1 segment per round trip and thus the aggregate window increases by n segments. Now, let us introduce a variant CA algorithm. Suppose that the aggregate window opens by m segments per round trip (Fig. 11-2). m could be a constant, for instance 1, or a variable such as $n/2$. Therefore, it takes $W_n/2mn$ round trips for the aggregate window to increase by $W_n/2n$ packets. Considering the area of each cycle, we have:

$$\frac{W_n}{2mn}\frac{(2n-1)W_n}{2n}+\frac{1}{2}\frac{W_n}{2mn}\frac{W_n}{2n}=\frac{1}{p} \tag{11.4}$$

Solving for the aggregate window W_n we get:

$$W_n = \sqrt{8mn^2/(4n-1)p} \tag{11.5}$$

Substituting W_n into the bandwidth equation, we get the estimated throughput with n connections and the variant CA algorithm:

$$BW_n = \frac{(1/p)*MSS}{(W_n/2mn)*RTT} = \sqrt{\frac{m(4n-1)}{2p}} \times \frac{MSS}{RTT} \tag{11.6}$$

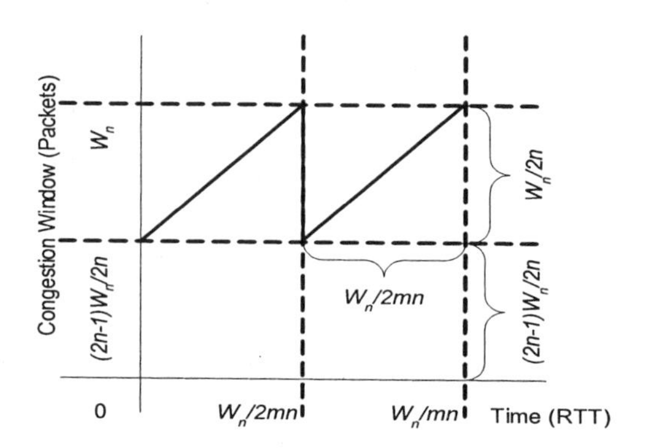

Figure 11-2. The evolution of the aggregate window with n virtual connections (Wn<Wm).

4.2 Case of $W_n >= W_m$

Under a given packet loss rate, as the number of connections increases the aggregate window grows and finally reaches the maximum window which is allowed by the link without inducing a queue (Fig. 11-3). Let the value of the maximum window be W_m packets and assume that the aggregate window, W_n, does not grow any more when the maximum window (W_m) is reached so that the RTT can be constant. Here we introduce a ratio, r:

$$r = W_m / W_1 \tag{11.7}$$

From Eq. (11.2) and Eq. (11.7), we have:

$$1/p = 3W_1^2/8 = 3W_m^2/8r^2 \tag{11.8}$$

$$W_m = \sqrt{8r^2/3p} \tag{11.9}$$

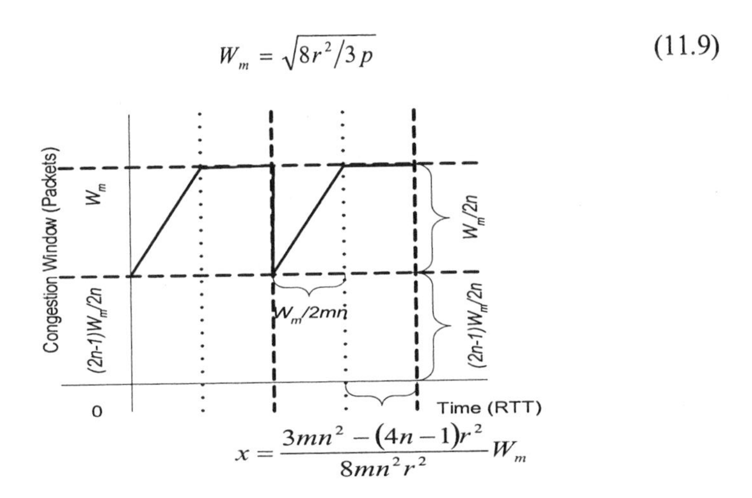

Figure 11-3. The evolution of the aggregate window with n virtual connections (Wn>=Wm).

Similar to the idea in Fig. 11-2, with the variant Congestion Avoidance algorithm, it takes $W_m/2mn$ round trips for the aggregate window to increase by $W_m/2n$ packets to reach W_m. After that, the window stays constant until a packet loss happens and is reduced by $W_m/2n$ to $(2n\text{-}1)W_m/2n$. Suppose this procedure takes x round trips. We have:

$$\frac{W_m}{2mn}\frac{(2n-1)W_m}{2n} + \frac{1}{2}\frac{W_m}{2mn}\frac{W_m}{2n} + xW_m = \frac{1}{p} \tag{11.10}$$

Substituting (11.8) in (11.10), we can get x:

$$x = \left[3mn^2 - (4n-1)r^2\right]W_m/8mn^2r^2 \tag{11.11}$$

Substituting x and W_m into the bandwidth equation, we get the estimated throughput with n connections and the variant CA algorithm:

$$BW_n = \frac{(1/p)*MSS}{\left(\frac{W_m}{2mn} + \frac{3mn^2-(4n-1)r^2}{8mn^2r^2}W_m\right)*RTT} = \frac{2mn^2r\sqrt{6}}{(3mn^2+r^2)\sqrt{p}} \times \frac{MSS}{RTT} \tag{11.12}$$

4.3 Bandwidth Estimation

From Eq. (11.6) and Eq. (11.12), if the aggregate window increases by m segments per round trip the estimated throughput based on n virtual connections is:

$$BW_n = \begin{cases} \sqrt{\dfrac{m(4n-1)}{2p}} \times \dfrac{MSS}{RTT}, & W_m > W_n \\ \dfrac{2n^2 mr\sqrt{6}}{(3n^2 m + r^2)\sqrt{p}} \times \dfrac{MSS}{RTT}, & W_m <= W_n \end{cases} \tag{11.13}$$

5. IMPACT OF CONGESTION AVOIDANCE

In this section, we describe our simulation model and simulation experiments designed to study the impact of CA on Parallel TCP. The simulations use LBL NS-2 and are carried out on the topology shown in Fig. 11-4, in which a TCP sender and receiver are connected by a single link. Table 11-1 is a concise summary of the system parameters selected for the simulations. This single link topology is far too simple to model the real complexity of an end-to-end connection through the Internet, however it enables us to explore the properties of the performance model.

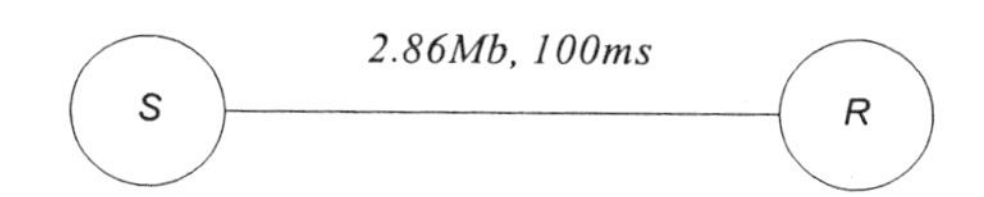

Figure 11-4. Simulation topology.

Table 11-1. Simulation parameters.

TCP Scheme	Reno TCP
Link Capacity	2.86Mb
Link Delay	100ms
RTT	202ms
Packet Size (MSS)	1,000bytes
Buffer Size	Unlimited
Error Distribution	Uniform
Simulation Time	1,000s

5.1 Static CA

Figs. 11-5 ~ 13 show throughput as a function of the number of connections under a variety of packet loss rates. Simul and Simu2 in the figures are the results of conducted simulations. Simul allows each individual connection to adjust its window according to its individual Congestion Control procedure. Simu2 measures the aggregate window and divides it by the number of connections to get an average window, which is then allocated to all the virtual connections. That is, immediately after a measurement of the aggregate window and window allocation, all the virtual connections have the same window.

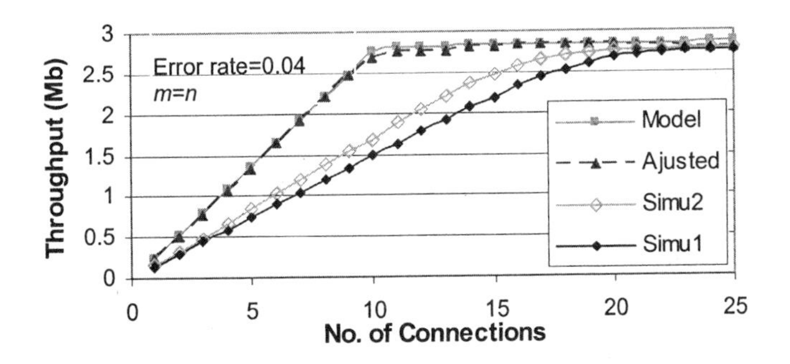

Figure 11-5. Throughput vs. no. of connections.

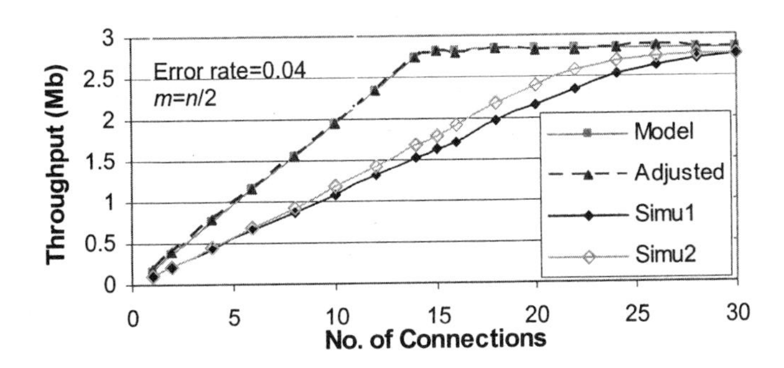

Figure 11-6. Throughput vs. no. of connections.

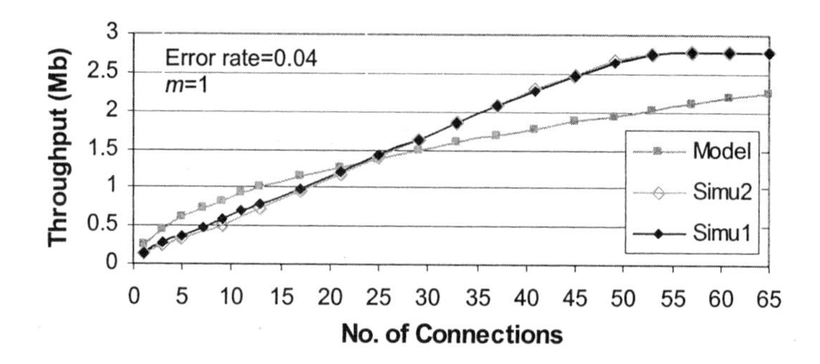

Figure 11-7. Throughput vs. no. of connections.

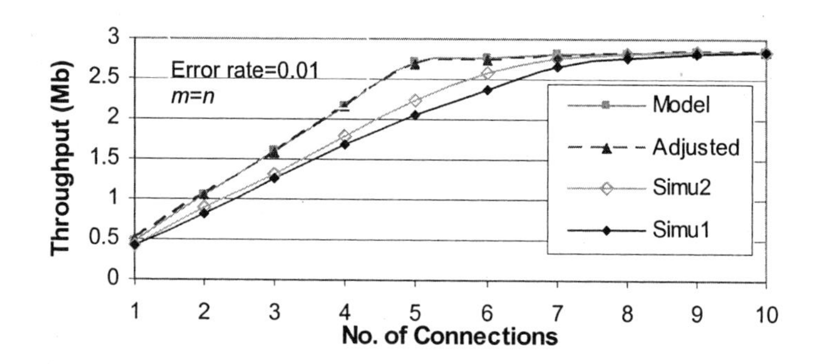

Figure 11-8. Throughput vs. no. of connections.

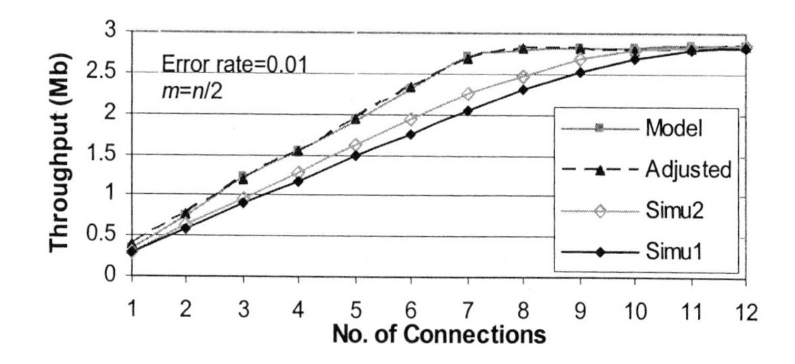

Figure 11-9. Throughput vs. no. of connections.

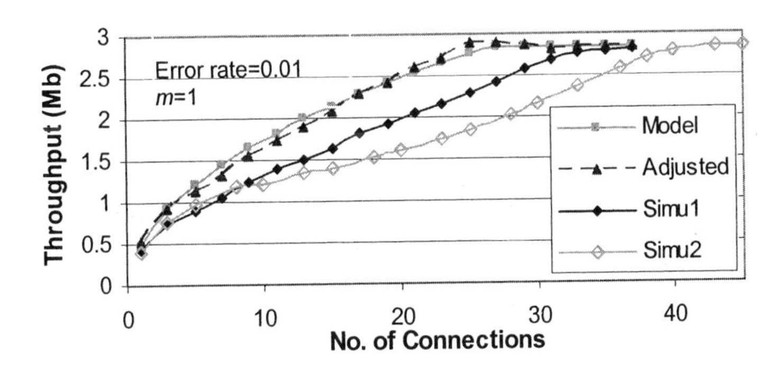

Figure 11-10. Throughput vs. no. of connections.

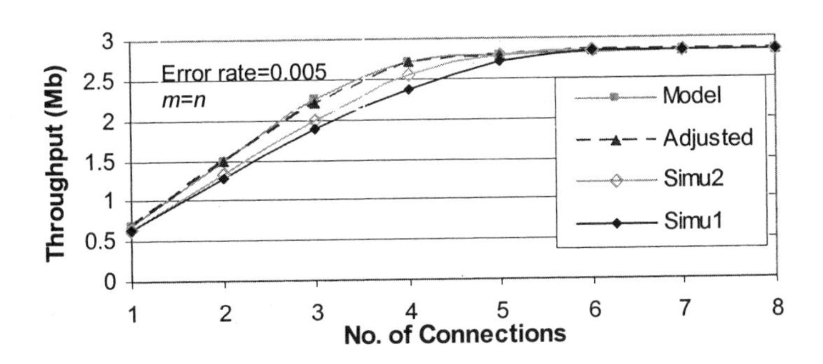

Figure 11-11. Throughput vs. no. of connections.

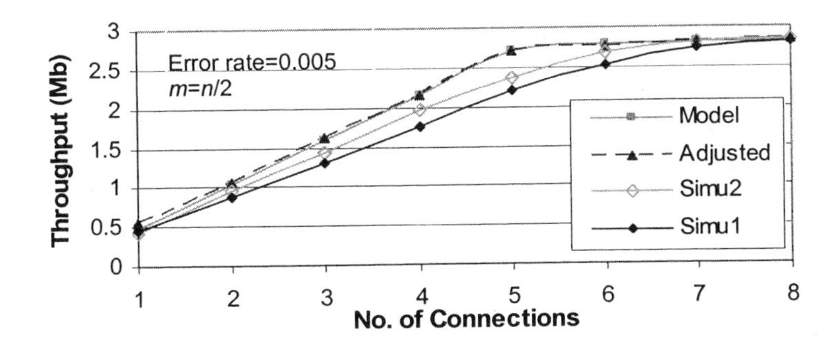

Figure 11-12. Throughput vs. no. of connections.

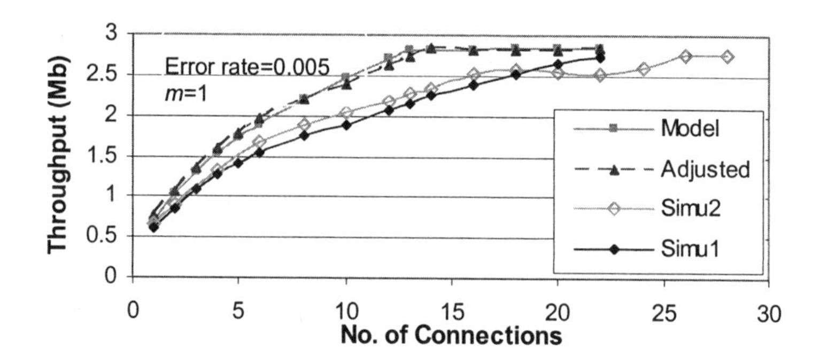

Figure 11-13. Throughput vs. no. of connections.

Figs. 11-5 ~ 7 show the simulations under a packet loss rate of 0.04. In Fig. 11-5 and 6, *m* (the increment of the aggregate window per round trip) in Eq. (11.13) is a variable which is a linear function of the number of connections, *n*. According to the model, with $m=n/2$ or *n*, throughput increases almost linearly with the number of connections in use as long as the aggregate window (W_n) is less than the maximum available window (W_m). It is reflected by the simulations in Fig. 11-5/6 as well as Fig. 11-8/9 (packet loss rate 0.01) and Fig. 11-11/12 (packet loss rate 0.005). These figures also show that Simu2 can give a better performance than Simu1. However, we notice in these figures that there is a gap between modeled throughput (Model) and measured throughput (Simu1 and Simu2). The gap increases with the packet loss rate. This is understandable as the assumptions underlying the model may not be satisfied under a high packet loss rate. However, the close match between the adjusted throughput and the modeled throughput proves that high packet loss rate does not significantly affect the accuracy of the model as explained below.

The adjusted throughput (BW_a) is achieved by introducing a coefficient for the measured throughput (BW_m) that can be either Simu1 or Simu2. The BW_a presented in the figures is based on Simu1.

$$BW_a = \begin{cases} BW_m * \alpha, & W_n < W_m \\ BW_m * \alpha^{(n-\eta+2)^{-[\mu+(n-\eta)\kappa]}}, & W_n > W_m \end{cases} \quad (11.14)$$

where α , η , μ and κ are a constant and *n* is the number of virtual connections. η is the benchmark number of connections where the W_n is equal to or just larger than W_m. α , μ and κ are experimentally selected according to the individual packet loss rates (Table 11-2). It shows that α and the gap mentioned before increase with packet loss rate. However, an appropriate selection of α , μ and κ can give a close match between the adjusted throughput and the modeled throughput.

Table 11-2. Selected values for α , μ and κ .

Error Rate	0.04		0.01		0.005	
	m=n	*m=n*/2	*m=n*	*m=n*/2	*m=n*	*m=n*/2
α	1.82	1.8	1.28	1.32	1.2	1.23
μ	0.1	0.11	1.0	0.5	0.4	0.0
κ	0.1	0.6	1.0	0.4	1.4	0.8
η	11	15	6	8	4	5

In Fig. 11-7, 10 and 13, *m* is a constant. For standard CA, the congestion window increases by 1 segment per round trip. With *m=n*, if there are a large number of virtual connections Parallel TCP connections would steal bandwidth from traditional TCP connections in the network, especially when random packet losses are present [27]. Therefore, we choose *m*=1 to match the standard CA. Fig. 11-7 shows that when the packet loss rate equals 0.04, Simu2 cannot give a better performance than Simu1 and the model is not suitable. The opposite is observed when *m=n*/2 or *n*. Compared to Fig. 11-5 and 6, where 19 and 26 connections are needed to achieve a throughput around 2.6Mb, in Fig. 11-7 it requires 49 connections. When the packet loss rate equals 0.01 (Fig. 11-10), adjusted Simu1 becomes consistent with the model but Simu2 still can not match the model and presents a smaller throughput than Simu1 after 8 connections. When the packet loss rate falls to 0.005 (Fig. 11-13), the adjusted Simu1 matches the model very well and in most cases Simu2 can give a larger throughput than Simu1, provided the number of connections is smaller than 19.

5.2 Dynamic CA

The purpose of this section is to demonstrate the impact of Dynamic CA using a fixed number of connections (DynCA) on TCP performance. This will be compared to the operations of Fixed Number of Connections (FixNC), Dynamic Number of Connections (DynNC) and Modified CA using a single connection (ModCA). These four approaches are discussed below. The following experiments have a packet loss rate of 0.005.

FixNC is the approach we used in Section 5.1 for Simu2 when *m=n*. That is, each individual virtual connection adopts the standard CA algorithm (the aggregate window opens by *n* packets, *n* being the number of connections) and their congestion window is centrally controlled.

DynNC is similar to FixNC, but uses an adaptive algorithm similar to [17] to control the number of connections based on changes in the observed

RTT. The algorithm introduces two RTT thresholds, *Low_RTT* and *High_RTT*. When the observed RTT exceeds *High_RTT*, one of the connections is closed, reducing the aggregate window by $1/n$. In [17], half of the connections are closed. We do not adopt this method because, similar to TCP Vegas where RTT is also used to detect the incipient stages of congestion, halving the number of connections means halving the aggregate window and therefore is too dramatic in the early stage of a potential congestion. When the RTT is below *Low_RTT*, DynNC adds one more connection. The new connection starts with Congestion Avoidance instead of Slow Start. If the RTT is between *Low_RTT* and *High_RTT*, the number of connections is not adjusted. Similar to FixNC, each individual virtual connection adopts the standard CA algorithm.

As with DynNC, DynCA uses the RTT-based algorithm to adaptively adjust the CA algorithm instead of the number of connections. When the RTT is above *High_RTT*, the increment in the CA algorithm is set to minus 1.0. This means that the aggregate window is shrunk by n packets per RTT. When the RTT falls below *Low_RTT*, the increment is set to positive 1.0. That is, the aggregate window opens by n packets per RTT. If the RTT is between *Low_RTT* and *High_RTT*, the increment is set to 0. It should be noticed that with more connections the aggregate window opens faster, but closes faster as well.

ModCA mimics the operation of parallel connections without introducing multiple connections. It therefore has similar properties to [12], [13] and [15]. If an operation of n parallel connections is emulated, ModCA sets the CA increment to n packets and the congestion window is cut down by $1/2n$ when a packet loss is detected. Note that if $n=1$ or there are no packet losses, ModCA functions as one or more standard TCP connections. The major difference between ModCA and FixNC is that FixNC is based on multiple connections which allow self-curing.

Figs. 11-14 and 11-15 compare between DynCA, FixNC and ModCA in terms of throughput and RTT. For DynCA, the *Low_RTT* and *High_RTT* are 204ms and 210ms, respectively. DynCA outperforms FixNC on RTT at the cost of slightly lower throughput. Notice in these two figures that after 6 connections DynCA basically does not increase its throughput, which is around 2.52Mb/s, even though the maximum throughput is potentially 2.86Mb/s. The reason for this is that, with more connections the aggregate window opens faster but closes faster as well. With 5 connections, FixNC seems to have relatively the best performance: an average throughput of 2.77Mb/s and an average RTT of 250.39ms, compared to the 2.40Mb/s and 204.97ms achieved by DynCA. It can be argued that the benefits of DynCA on RTT performance outweigh its performance loss on throughput. ModCA has the lowest throughput because of the lack of self-curing. Its performance on RTT is better than FixNC because of lower throughput, but below DynCA's. It is expected that if the packet loss rate increases, compared to DynCA and FixNC, it will have a worse performance on throughput, and if the packet loss rate falls it will behave more like FixNC.

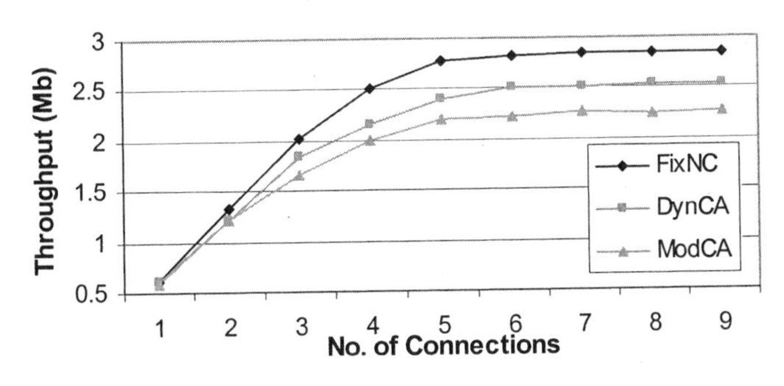

Figure 11-14. Comparison between DynCA, FixNC and ModCA on throughput.

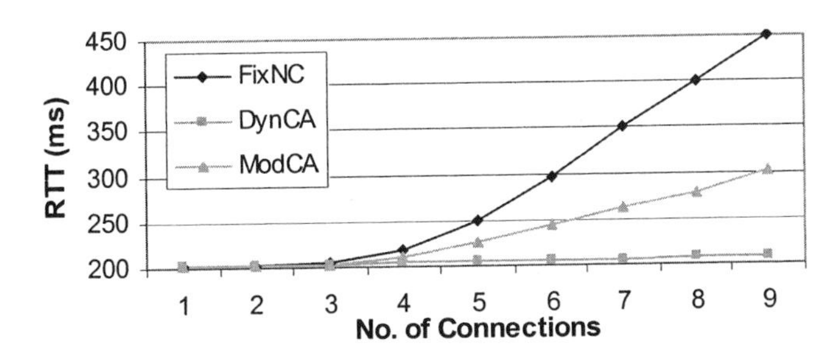

Figure 11-15. Comparison between DynCA, FixNC and ModCA on RTT.

If there are too many connections FixNC can create congestion (Fig. 11-15). In contrast, DynNC has an adaptive algorithm to control the number of connections being used over time. Figs. 11-16 and 11-17 show the frequency of different number of connections used during a DynNC simulation run where the *Low_RTT/High_RTT* pairs are 204ms/210ms and 208ms/220ms respectively. The parameter, *Jump*, is the number of jumps which happen when the number of connections is adjusted. The jump happens almost twice per second in Fig. -16 and 3 times per second in Fig. 11-17. This is bad for single-socket operations. Fig. 11-16 shows that 5 connections are most frequently used during the transmission. Referring to Fig. 11-14 and 11-15, it is observed that with 5 connections the average throughput (2.40Mb/s) and

RTT (204.97ms) achieved by DynCA outperform the values (2.14Mb/s and 207.87ms) achieved by DynNC. Furthermore, as long as there are more than 3 connections DynCA performs better than DynNC.

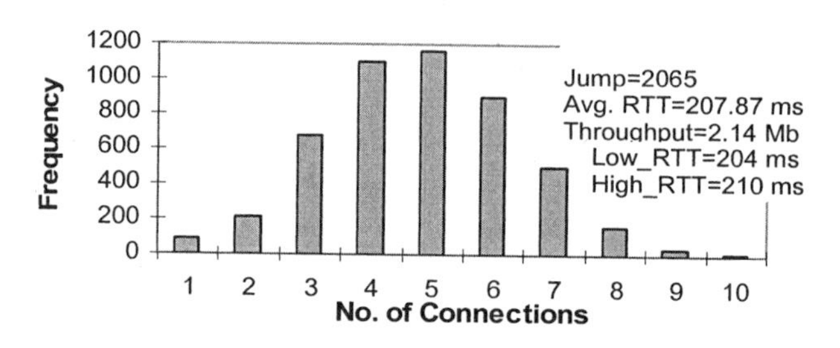

Figure 11-16. Frequency of different number of connections with DynNC

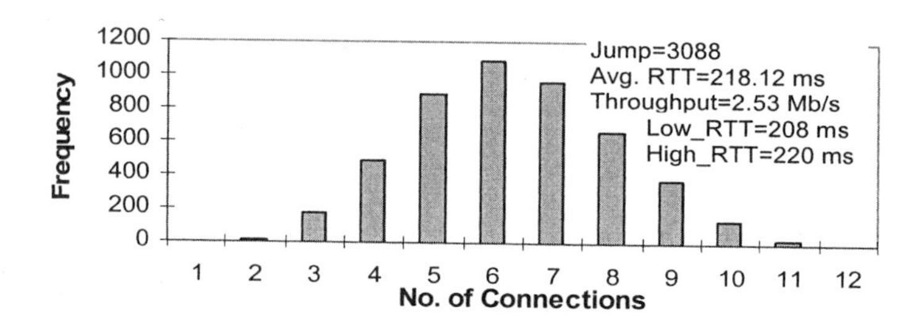

Figure 11-17. Frequency of different number of connections with DynNC

When the *Low_RTT-High_RTT* range is double from 6ms (210ms-204ms) to 12ms (220ms-208ms), the most frequently used number of connections changes from 5 to 6, as can be seen in Figs. 11-16 and 11-17. The number of jumps also increases from 2065 to 3088. It is surprising that Fig. 11-17 has more jumps because it has a doubled *Low_RTT-High_RTT* range. For the simulation with *Low_RTT/High_RTT* at 204ms/216ms, which also has a *Low_RTT-High_RTT* range of 12ms with a lower position in the time scale, the jump, average RTT and throughput are 1808, 210.6ms and 2.32Mb/s. This means that the number of jumps is not only related to the *Low_RTT-High_RTT* range but also its position in the time scale.

Figs. 11-18 ~ 21 are the throughput and RTT traces obtained from DynNC and DynCA simulations respectively, with the *Low_RTT/High_RTT* pair set to 208ms/220ms. The DynCA simulation is carried out with 6 connections. It is not surprising that both approaches have a throughput gain at the cost of a longer RTT. Once again, DynCA outperforms DynNC in terms of average throughput, average RTT and their corresponding standard deviation.

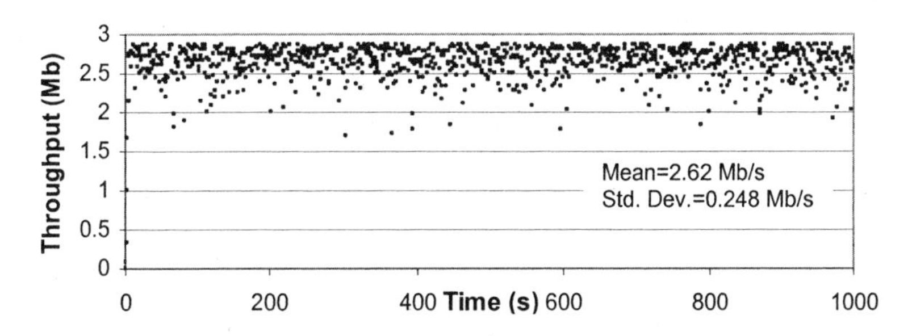

Figure 11-18. Throughput trace of DynCA with 6 connections

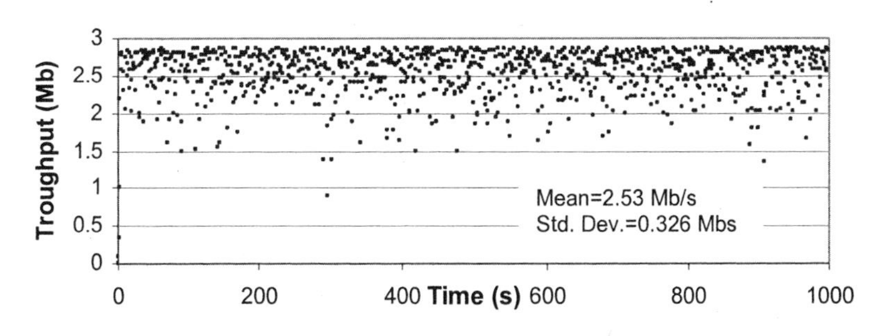

Figure 11-19. Throughput trace of DynNC

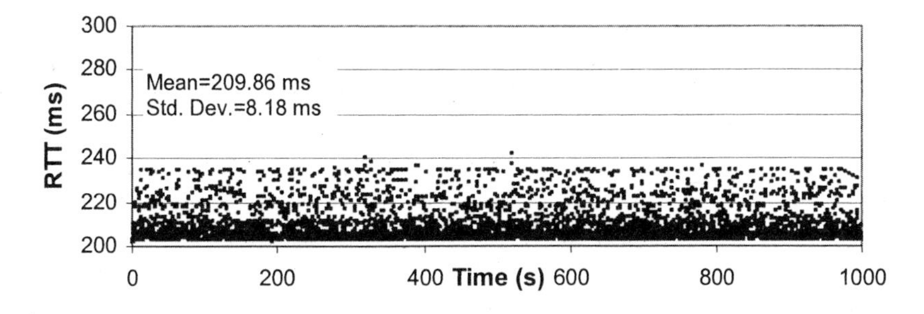

Figure 11-20. RTT trace of DynCA with 6 connections

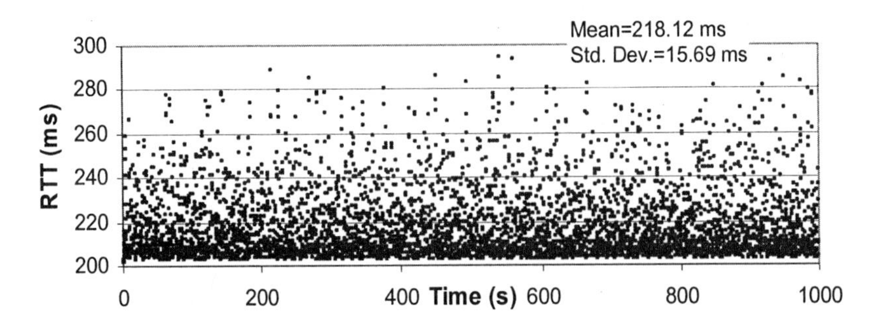

Figure 11-21. RTT trace of DynNC

The experiments shown in this section indicate that it is possible to adopt an adaptive CA algorithm with a fixed number of connections to reflect network dynamics and achieve reasonable performance.

6. CONCLUSIONS

In this chapter, we have presented the motivation behind Parallel TCP, which is a TCP extension to suit heterogeneous networks with wired and wireless (possibly multihop) links. Both the features of the protocol and an analytical performance model of Parallel TCP have been described. The model takes Congestion Avoidance (CA) into account. The model explains how Parallel TCP can improve throughput performance and shows the factors that affect the performance improvement, such as the number of connections, packet loss rate, CA, etc. A simulation model of Parallel TCP was created for a simple network topology. Simulations were carried out under a variety of CA schemes as well as different packet loss rates. Although the difference between modeled and measured (by simulation) throughput is significant and becomes more significant with the increase of the packet loss rate, there is a close match between modeled and adjusted throughput. The adjusted throughput is obtained by introducing a coefficient for the measured throughput.

It has been shown that if the virtual connections adopt the standard CA algorithm or similar schemes in which the individual congestion window opens by a constant number of segments per round trip, the analytically modeled throughput matches the adjusted throughput well even under high packet loss rates (see Figs. 11-5, 6, 8, 9, 11 and 12 for detail). However, if the aggregate window opens by a constant number of segments the analytical model fails if the packet loss rate is high (Fig. 11-7). The CA algorithm used in the former case is more aggressive than the one in the latter case. We have also shown that in the former case an improved throughput can be achieved by allocating the same window to all the virtual connections. However, it may not always be applicable in the latter case.

With the increase of the packet loss rate, more connections and a more aggressive CA algorithm are preferable to utilize the bandwidth. However, when congestion occurs, it is needed to either reduce the number of connections or adopt a less aggressive CA algorithm, or both. For the single-socket operations, it is preferred to maintain a fixed number of connections. When the random packet loss prevails maintaining a certain number of connections can take advantage of self-curing (the bandwidth released from a disrupted connection can be used by the other connections).

The experiments indicate that it is possible to dynamically adjust the CA algorithm in a fixed number of connections instead of changing the number of connections to reflect network dynamics and thus to achieve a reasonable performance. Moreover, we have shown the advantage of self-curing based on multiple connections when the random packet loss is present.

ACKNOWLEDGMENTS

We are grateful to Ryan Wishart for his careful reading and valuable comments on this chapter.

REFERENCES

1. G. Xylomenos, G. Polyzos, P. Mahonen and M. Saaranen, "TCP Per-formance Issues over Wireless Links", *IEEE Comm. Mag.*, April, 2001.
2. J. Lee, D. Gunter, B. Tierney, W. Allock, J. Bester, J. Bresnahan and S. Tecke, "Applied Techniques for High Bandwidth Data Transfers across Wide Area Networks", LBNL-46269, 2000.
3. J. Bolot, "Characterizing End-to-End packet delay and loss in the Internet", *J. of High Speed Networks*, 2(3), 1993.
4. T. Lakshman and U. Madhow, "The performance of TCP/IP for networks with high bandwidth-delay products and random loss," *IEEE/ACM Transactions on Networking*, 5(3), 1997.
5. V. Jacobson, "Congestion avoidance and control", *ACM SIGCOMM*, August 1988.
6. S. Floyd, "A Report on Recent Developments in TCP Congestion Control", *IEEE Comm. Mag.*, April, 2001.
7. V. Tsaoussidis and I. Matta, "Open issues on TCP for mobile com-puting", *J. of Wireless Comm. & Mobile Computing*, 2(2), 2002.
8. C. Perkins, ed., "IP Mobility Support for IPv4", *RFC3220*, Jan. 2002.
9. L. Brakmo and L. Peterson, "TCP Vegas: End to End Congestion Avoidance on a Global Internet", *IEEE J. Sel. Areas Comm.*, 13 (8), 1995.

10. W. Xu, A. Qureshi and K. Sarkies, "a Novel TCP Congestion Control Scheme and Its Performance Evaluation", *IEE Proc. Comm.*, 149(4), Aug. 2002.
11. Q. Fu and L. White, "The Impact of Background Traffic on TCP Performance over Indirect and Direct Routing", *ICCS*, 2002.
12. Tom Kelly, "Scalable TCP: Improving performance in highspeed wide area networks", *Computer Communication Review* 32(2), April 2003.
13. Sally Floyd, "HighSpeed TCP for Large Congestion Windows", *RFC 3649*, Dec. 2003.
14. Dina Katabi, Mark Handley, and Chalrie Rohrs, "Congestion Control for High Bandwidth-Delay Product Networks", *ACM SIGCOMM*, 2002.
15. Cheng Jin, David X. Wei and Steven H. Low, "FAST TCP: motivation, architecture, algorithms, performance", *IEEE INFOCOM*, March 2004
16. K. K. Ramakrishnan and S. Floyd, "Proposal to add explicit congestion notification (ecn) to IP", *RFC 2481*, Jan. 1999.
17. M. Allman, H. Kruse and S. Ostermann, "An Application-Level Solution to TCP's Satellite Inefficiencies", *Proc. WOSBIS*, Nov. 1996.
18. H. Balakrishnan, H. Rahul, and S. Seshan, "An Integrated Congestion Management Architecture for Internet Hosts", *ACM SIGCOMM*, Sept. 1999.
19. L. Eggert, J. Heidemann, and J. Touch, "Effects of Ensemble–TCP", *ACM Computer Communication Review*, 30(1), pp. 15-29, Jan. 2000.
20. M. Allman, V. Paxson, and W. Stevens, "TCP Congestion Control", *RFC 2581*, April 1999.
21. D. Iannucci and J. Lekashman, "MFTP: Virtual TCP Window Scaling Using Multiple Connections", RND-92-002, NASA Ames Research Centre, Jan. 1992.
22. J. Hahn, "MFTP: Recent Enhancements and Performance Measurements", RND-94-006, NASA Ames Research Centre, June 1994.
23. H.-Y. Hsieh and R. Sivakumar, "A Transport Layer Approach for Achieving Aggregate Bandwidths on Multi-homed Mobile Hosts", *MOBICOM*, Sept. 2002.
24. H. Sivakumar, S. Bailey, and R. Grossman, "PSockets: The case for application-level network striping for data intensive applications using high speed wide area networks", *Proc. IEEE Supercomputing*, 2000.
25. M. Mathis, J. Semke, J. Mahdavi, T. Ott. "The Macroscopic Behavior of the TCP Congestion Avoidance Algorithm", *Computer Comm. Review*, 27(3), July 1997.
26. T. Hacker, B. Athey and B. Noble, "The End- to-End Performance Effects of Parallel TCP Sockets on a Lossy Wide-Area Network", Proc. IPDPS, April 2002.
27. T. Hacker, B. Noble and B. Athey, "The Effects of Systemic Packet Loss on Aggregate TCP Flows", *Proc. IEEE Supercomputing*, 2002
28. A. Zanella, G. Procissi, M. Gerla, M. Sanadidi, "TCP Westwood: Analytic Model and Performance Evaluation", *Globecom*, Nov. 2001.
29. R. Stewart, Q. Xie, K. Morneault, C. Sharp, H. Schwarzbauer, T. Taylor, I. Rytina, M. Kalla, L. Zhang, and V. Paxson, "Stream Control Transmission Protocol", *RFC 2960*, October 2000.
30. Q. Fu and J. Indulska, "Parallel TCP for Error Prone Links", *RIUPEEEC*, 2003.
31. Q. Fu and J. Indulska, "Splitting TCP in Parallel for Error Prone Links", *ICICS-PCM*, 2003.
32. M. Allman, S. Floyd and C. Partridge, "Increasing TCP's Initial Window", *RFC 2414*, September 1998.
33. M. Allman, C. Hayes and S. Ostermann, "An Evaluation of TCP with Larger Initial Windows", *Computer Comm. Review*, 28(3), July 1998.

Chapter 12

PERFORMANCE ANALYSIS OF RELIABLE MULTICAST PROTOCOLS: A MESSAGE-BASED APPROACH

Julija Tovirac,[1] Weimin Zhang[2,1] and Sylvie Perreau[1]
[1] *Institute for Telecommunications Research*
University of South Australia
[2] *Defence Science and Technology Organisation*
Edinburgh, South Australia

Abstract: In this chapter we derive a mathematic model that analyses the performance of some generic reliable multicast protocols. Two error recovery mechanisms have been considered, namely a pure ARQ and a hybrid ARQ with ideal erasure codes. The novelty of our model dwells on its message-based approach rather than packet based calculations, as usually found in the literature. The model matches well with those reliable multicast protocols with aggregated acknowledgments for file and message deliveries. The expressions derived cater for receivers with non-identical channel loss distributions. The dynamics of the multicast group size are also studied. Simulations were conducted and the results agree well with the analysis.

Key words: Reliable Multicast, ARQ, Hybrid ARQ, Ideal Erasure FEC, P_Mul, Message-Based Protocol, Throughput

1. INTRODUCTION

An increasing number of applications is based on simultaneous dissemination of information to a group of receivers. The introduction of multicast into the network architecture promises many benefits to the network that serves such applications but it also opens many questions and challenges [1]. Using multicast to communicate to a group of receivers overall reduces the amount of traffic as compared with the establishment of equivalent number of multiple unicast connections. The main advantage of multicast communication is in overall minimisation of the number of transmissions via shared network paths. However, the provision of efficient scalable reliable multicast error recovery mechanisms is still one of the challenges. The presence of a multicast source in a local network can have a considerable impact on the local traffic dynamics and it is important to estimate the amount of generated and transmitted data.

Pingali et al studied the throughput performance of sender-based and receiver-based multicast protocols [2, 3]. A similar study on bandwidth efficiency with selective-repeat and go-back-n was presented in [4]. Levine et al [5] extended the work of Pingali et al on tree-based and ring-based protocols. Nonnenmacher et al [6] focused on the analysis of a multicast protocol with layered and integrated forward error control (FEC) coding. All of the above mentioned authors assumed an infinite packet stream when deriving expressions for the expected number of retransmissions. They then generalised their findings from a single packet in an infinite file. In our work we restrict this assumption to a file or a message consisting of a fixed number of information packets, L.

We derive a mathematical model that describes the error recovery process of reliable multicast protocols with two error recovery mechanisms. One of them is based on automatic repeat request (ARQ) only and the other is based on hybrid ARQ, which is an ARQ enhanced with an ideal FEC code and erasure decoding. The model is generic and should be applicable to many file based reliable multicast protocols, although it is developed from P_Mul [7], which is a message and file multicast protocol. Our goal, in this work, is to find out the average number of packets the sender has to transmit until every receiver recovers the file. We also study how the size of the multicast group evolves with each retransmission round.

The paper is organised as follows. We first introduce the network scenario and common parameters. Then, we describe models for both ARQ and hybrid ARQ with detailed derivations and reasoning behind formulae. Finally, we compare the theoretical performance with simulation results.

2. SCENARIO AND PARAMETERS

We consider a multicast group consisting of one sender S and G receivers, $R_1 \ldots R_G$, with G forward channels and G backward control channels, see Fig. 12-1.

All forward channels are mutually independent packet erasure channels. Generally the packet erasures are caused by packets being dropped due to either network congestion or detection of uncorrectable erroneous bits. Here

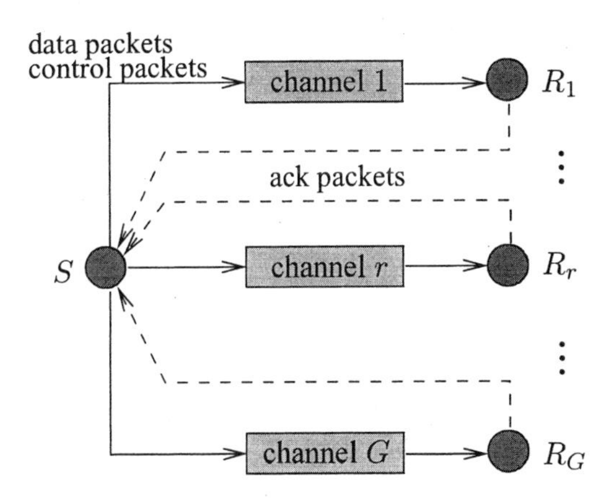

Figure 12-1. Simplified network model

we assume dominance of the latter factor: an error detection scheme discards packets that contain errors introduced at the physical layer. Channel independency is a pessimistic assumption taking into account that in practice receivers may experience certain degree of correlation between their losses. The error distributions for each channel do not have to be identical. The backward channels are assumed ideal for the reliable transfer of acknowledgment packets. From here we will refer to forward channels as the channels. Furthermore, we assume that the sender has sufficient processing power and memory to handle the responses from all receivers.

We consider the following scenario. The sender wishes to reliably multicast a file of size $l \times L$ bytes to each receiver in the group. The file is divided into L packets, each consisting of an h-byte header and an l-byte payload. Each multicast transmission round starts with an announcement from the sender, listing those relevant receivers. This is known as an Address_PDU packet in P_Mul. Data packets are then transmitted following the Address_PDU.

Depending on the protocol, the receivers respond with acknowledgments (Acks), listing which packets have been received, or with negative acknowledgments (NAcks), listing packets that have been missed. On reception of all Acks (or NAcks), the sender compiles a list of packets to be retransmitted in the following round. The receivers that acknowledged reception of all file packets are excluded from consideration in the following rounds. The group gradually shrinks in size and once all receivers have confirmed the file reception, the group size becomes 0 and the file recovery process is complete. We assume that none of the control packets, Ack, NAck or Address_PDU, experience any loss. Nevertheless, their influence in terms of propagation and transmission overhead is considered in the performance analysis (section 5.1).

The given scenario and initial conditions are the same for both ARQ and hybrid ARQ. The difference is in the way these schemes handle retransmissions. The Table 12-1 lists parameters and notations that are common to both schemes.

Table 12-1. Notations

r	The receiver and channel index, $1 \leq r \leq G$.
i	The ith retransmission round. $0 \leq i \leq \infty$, where $i = 0$ is the initial transmission, $i = 1$ is the first retransmission, etc.
L_i	The average number of packets sent by the sender in the ith round. $L_0 = L$, while the average of the total number of packets is $\sum_{i=0}^{\infty} L_i$.
G_i	The average size of the multicast group in the ith round, where $G_0 = G$.
$q_i^{(r)}$	The average number of packets that receiver r expects in the ith round of transmission, where $q_0^{(r)} = L \quad \forall r = 1 \ldots G$.
$P_E^{(r)}$	The packet erasure probability for channel r and receiver r. It depends on the channel bit error rate, ber, the distribution of bit errors (here assumed random) and the packet length, $l+h$.
$P_C^{(r)}$	The probability that a packet is error-free, e.i. $P_C^{(r)} = 1 - P_E^{(r)}$.

3. MULTICAST WITH ARQ

In this section we derive the expressions for L_i, q_i and G_i, for the multicast protocol with the ARQ error recovery scheme. We use the principles of mathematical induction – observe the behaviour through rounds $i = 0, 1, 2, \ldots$ and generalise it for an arbitrary round i.

The main feature of the reliable multicast ARQ is that the sender selectively retransmits the requested data until all receivers confirm the file reception. The initial conditions $(i=0)$ are given in section 2, Table 12-1. In addition to that, we introduce one more variable – the probability that an arbitrary data packet is confirmed by all receivers in the group in round i, $P_{C,i}^{(group)}$. This is the probability that a packet is *not* retransmitted in the following round. A packet is retransmitted in round $i+1$ if it has *never* been confirmed by *at least one* receiver in the group. The probability of this event is denoted $P_{E,i}^{(group)}$. Naturally, $\forall i$

$$P_{C,i}^{(group)} + P_{E,i}^{(group)} = 1 \tag{12.1}$$

and,

$$L_{i+1} = L_i \cdot P_{E,i}^{(group)}. \tag{12.2}$$

3.1 Calculating L_i for multicast with ARQ

In the initial transmission, $i=0$, the sender multicasts the entire file of L packets and waits for the group's response. At that time every receiver in the group expects to obtain $q_0=L$ packets. Due to the channel loss, each receiver on average gets $LP_C^{(r)}$ packets and loses $LP_E^{(r)}, \forall r=1\ldots G$, see Fig. 12-2. The destination nodes inform the sender on what has been received, and on average request $q_1^{(r)}$ packets,

$$q_1^{(r)} = LP_E^{(r)}. \tag{12.3}$$

Once all expected acknowledgments arrive, or a response timer expires, the

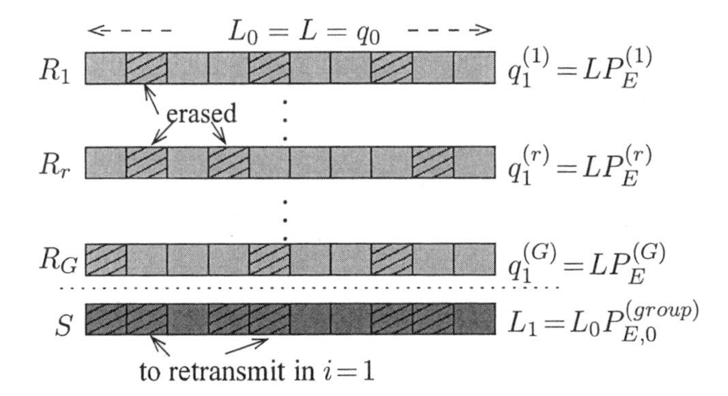

Figure 12-2. Sender and receivers states between the initial transmission ($i = 0$) and the first retransmission ($i = 1$)

sender goes through requests of all receivers and compiles a union list of all unconfirmed packet sequence numbers. Packets corresponding to the listed sequence numbers are retransmitted in the following round. The average number of retransmitted packets in $i=1$ is L_1 packets, $L_1 \leq L$. Now, the probability that a packet *is* successfully acknowledged by *all* receivers in the group in the round $i=0$, $P_{C,0}^{(group)}$, is

$$P_{C,0}^{(group)} = \prod_{r=1}^{G} P_C^{(r)} = \prod_{r=1}^{G}(1 - P_E^{(r)}). \tag{12.4}$$

Thus, the average number of packets that need to be retransmitted in the round $i=1$ is:

$$L_1 = L_0\, P_{E,0}^{(group)} = L\,(1 - \prod_{r=1}^{G} P_C^{(r)})\,. \tag{12.5}$$

The example given in Fig. 12-2 continues in Fig. 12-3. It shows the sender's and receivers' statuses after the first retransmission. It is clear that $q_i^{(r)} \le L_{i+1}$ always holds. This is because among those L_i packets retransmitted by the sender, some packets are requested by a receiver while others are repetitions of what have been already received. In other words, every receiver will consider the packets as either *irrelevant* – because they were received in the past, or *relevant* – lost previously and requested for retransmission. In rounds $i>1$, for a particular receiver the sender only needs to retransmit packets that are both erased and relevant, see Fig. 12-3.

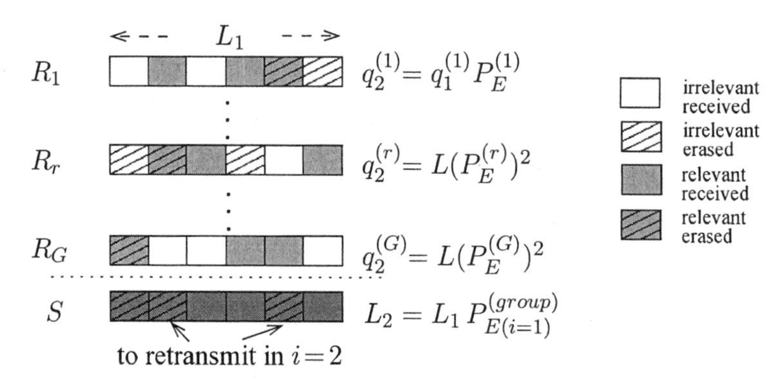

Figure 12-3. Sender and receivers states between the first retransmission ($i = 1$) and the second retransmission ($i = 2$)

What is the probability that a packet is relevant and erased?
The probability that a packet is relevant to the receiver r in the round i is

$$P_i^{(r)}(\text{pk relevant}) = \frac{q_i^{(r)}}{L_i}, \tag{12.6}$$

and irrelevant,

$$P_i^{(r)}(\text{pk irrelevant}) = 1 - P_i^{(r)}(\text{pk relevant}).$$

The relevance and erroneousness are independent events,

$$\begin{aligned} Pr(\text{pk relevant}|\text{erased}) &= Pr(\text{relevant}), \\ Pr(\text{pk erased}|\text{relevant}) &= Pr(\text{erased}). \end{aligned}$$

Therefore, a probability that a packet is relevant *and* erased is

$$\begin{aligned} P_i^{(r)}(\text{pk relevant} \cap \text{erased}) &= P_i^{(r)}(\text{pk relevant})\; P_E^{(r)}, \\ &= \frac{q_i^{(r)}}{L_i} P_E^{(r)}. \end{aligned} \tag{12.7}$$

Using mathematical induction it can be shown that $\forall i > 0$

$$q_i^{(r)} = P_E^{(r)}\, q_{i-1}^{(r)} = L(P_E^{(r)})^i. \tag{12.8}$$

Eq. (12.7) then becomes

$$P_i^{(r)}(\text{pk relevant} \cap \text{erased}) = \frac{(P_E^{(r)})^{i+1}}{L_i} L. \tag{12.9}$$

It allows us to derive the probability that a packet is successfully acknowledged by all receivers in the group in round i, $P_{C,i}^{(group)}$. That is, the probability that a packet will *not* have to be transmitted to the group again,

$$P_{C,i}^{(group)} = \prod_{r=1}^{G} (1 - P_i^{(r)}(\text{pk relevant} \cap \text{erased}))\,. \tag{12.10}$$

It can be shown that (12.4) is a special case of (12.10). Starting from (12.5) and combining (12.9) and (12.10), we can establish a general relationship between L_i and L_{i+1},

$$\begin{aligned} L_{i+1} = L_i\, P_{E(i)}^{(group)} &= L_i\,(1 - P_{C,i}^{(group)})\,, \\ &= L_i\,(1 - \prod_{r=1}^{G}(1 - \frac{q_i^{(r)}}{L_i} P_E^{(r)})). \end{aligned} \tag{12.11}$$

The general expression for L_i is then given as

$$L_i = L_{i-1}\,(1 - \prod_{r=1}^{G}(1 - L\frac{(P_E^{(r)})^i}{L_{i-1}})), \quad \forall i > 0. \tag{12.12}$$

We evaluate this expression in section 5 and compare it with results obtained using simulations.

3.2 Calculating G_i for multicast with ARQ

What happens to the average size of the multicast group? In practice, receivers that acknowledge the file reception remain passive members of the multicast group. They are ignored from consideration when assembling a new list of retransmitted packets for the following round. Therefore the average size of the multicast group shrinks from G to some average size G_1 in the first round. The group continues to reduce its size until it becomes 0, $G_i = 0$, at some round $i > 0$. Receiver r is excluded from the group with the probability that it has recovered the entire file in a particular round i. This probability is denoted $P_{CL,i}^{(r)}$ for some $i \geq 0$, and it can be expressed as

$$P_{CL,i}^{(r)} = (P_C^{(r)})^{q_i^{(r)}}. \tag{12.13}$$

It practically means that every packet in the file has been received. The probability that a receiver has at least one packet erased and it will remain in the multicast group in the following round, round $i+1$ is then

$$P_{EL,i}^{(r)} = 1 - P_{CL,i}^{(r)}. \tag{12.14}$$

Furthermore, $\forall i > 0$, $p_0^{(i)}, p_1^{(i)}, \ldots, p_G^{(i)}$ are probabilities that, respectively, a number of $0, 1, \ldots G$ receivers were unsuccessful in the $(i-1)^{th}$ round and will remain in the group in the ith round,

$$p_0^{(i)} = \prod_{r=1}^{G} P_{CL,i-1}^{(r)}, \tag{12.15}$$

$$p_1^{(i)} = \sum_{r=1}^{G} (P_{EL,i-1}^{(r)} \prod_{\forall r' \neq r} P_{CL,i-1}^{(r')}),$$

$$\vdots$$

$$p_g^{(i)} = \sum_{\forall \mathcal{G}_g \subset \mathcal{G}} (\prod_{\forall r \in \mathcal{G}_g} P_{EL,i-1}^{(r)} \prod_{\forall r' \in \mathcal{G}'_g} P_{CL,i-1}^{(r')}), \tag{12.16}$$

$$\vdots$$

$$p_G^{(i)} = \prod_{r=1}^{G} P_{EL,i-1}^{(r)}. \tag{12.17}$$

Eq. (12.16) defines the probability that g receivers remain active in the group in round i, $p_g^{(i)}$. There, $\mathcal{G}_g$ is a $g{-}subset$, i.e. a subset of $\mathcal{G}$ that contains exactly g elements, while $\mathcal{G}'_g = \mathcal{G} \setminus \mathcal{G}_g$, i.e. the complement of $\mathcal{G}_g$. Eq. (12.15) – (12.17) completely describe a probability density function whose average is the number of the remaining receivers in the group in round i, G_i,

$$G_i = \sum_{g=0}^{G} g\, p_g^{(i)}. \tag{12.18}$$

In Section 5 we will show an example of the network with i.i.d. channels. All expressions given in the last two sections have much simpler forms. A comparison of theoretical and simulation results will also be given there.

4. MULTICAST WITH IDEAL HYBRID ARQ

By ideal hybrid ARQ we mean a reliable multicast protocol enhanced by a perfectly efficient rateless erasure correcting code, i.e. a code with rate $\rightarrow 0$. Rateless codes, such as LT [8] and on-line codes [9], have a potential to generate unlimited number of unique parity (repair) packets, but exhibit a certain degree of decoding inefficiency [10] which is particularly noticeable in decoding of small files.

We assume that the sender transmits the file in the first round only, and new unique parities in all of the following rounds. It never multicasts any of the previously transmitted or reported lost packets, because it has access to an infinite pool of repair (parity) packets. At the receiver side, the decoding is assumed to be efficient: any new parity packet can replace any erased packet. Thus a receiver can recover the file as soon as it obtains exactly any L unique packets. The number of transmitted parity packets depends only on the worst case receiver request for that round. Acknowledgments do not need to be explicit with regard to packet IDs – it is sufficient for each receiver to report the number of packets it requires.

4.1 Protocol description

In this section we describe the interaction between the sender and an arbitrary receiver r, in round i:

- Assume the sender multicasts L_i packets on average, in round i;
- Assume, the receiver r requires any $q_i^{(r)} \leq L_i$ packets on average to complete the file reception, in round i;
- The receiver actively observes all L_i packets and reports to the sender its loss (or reception) of $L_i\, P_E^{(r)}$ (or $L_i\, P_C^{(r)}$) packets, on average. The probability that the receiver r reports j packets being lost if it expects

$q_i^{(r)}$ while L_i was transmitted is denoted $\Pr[X_r = j, i] = P_{E,i,j}^{(r)}$, and expands into

$$P_{E,i,j}^{(r)} = \begin{cases} \sum_{k=0}^{L_i - q_i^{(r)}} \binom{L_i}{k} (P_E^{(r)})^k \, (P_C^{(r)})^{L_i - k}, & j=0 \\ 0, & j \in (0, \; L_i - q_i^{(r)}] \\ \binom{L_i}{j} (P_E^{(r)})^j \, (P_C^{(r)})^{L_i - j}, & j \in (L_i - q_i^{(r)}, L_i\,]; \end{cases} \quad (12.19)$$

- If the receiver obtains at least $q_i^{(r)}$ out of L_i packets, it sends back a final Ack, and it is excluded from further consideration.
- Otherwise, the receiver's reported loss is taken into account when making decisions for $(i+1)^{th}$ round.
- In the $(i+1)^{th}$ round the sender on average transmits

$$L_{i+1} = E[\max(P_E^{(1)} L_i, \ldots, P_E^{(G)} L_i)] \quad \text{packets.}$$

4.2 Calculating L_i for multicast with ideal hybrid ARQ

The derivation begins with a similar approach as in section 3.1. Fig. 12-4 depicts the receivers' buffers in the course of the initial transmission, $i = 0$.

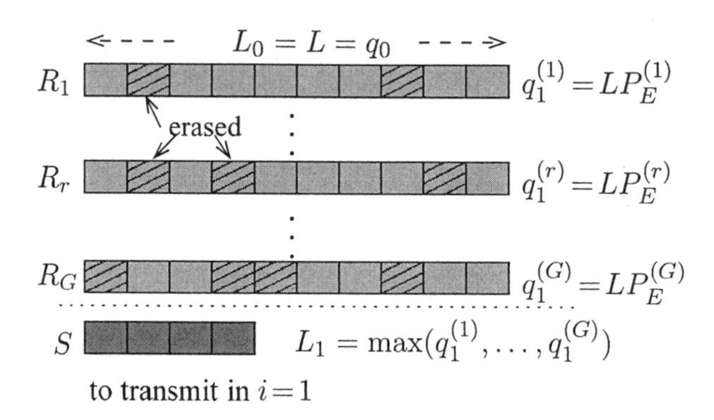

Figure 12-4. The sender and receivers states between first transmission $(i = 0)$ and first retransmission $(i=1)$

Each receiver on average loses $L\,P_E^{(r)}$ packets and reports $L\,P_E^{(r)}$ missed packets. The sender observes all responses and in the following round transmits L_1 parity packets on average. L_1 is estimated as an expectation of the worst case receiver request and equals to $L_1 = E[\max(q_1^{(1)}, \ldots, q_1^{(G)})]$ parity packets. In the given example (Fig. 12-4) the worst case receiver is R_G requesting 4 packets, therefore $L_1 = 4$. In Fig. 12-5 the example continues for the second round of transmissions (i.e. first round of retransmissions, $i = 1$). It can be noticed that now R_1 has received 2 packets, and hence completed the file reception because it required only 2 packets. Its loss of the other two packets is not important any more. Receiver G expected 4 packets to complete the file, but it has obtained 3, therefore it reports 1 lost packet. Receiver r lost 2 and received 2 packets, so it reports 2 lost packets, although it really needs 1. From the example illustrated in Fig. 12-4 and 12-5 it can be concluded that the average number or expected packets by each receiver is:

$$q_{i+1}^{(r)} = q_i^{(r)} - P_C^{(r)} L_i. \tag{12.20}$$

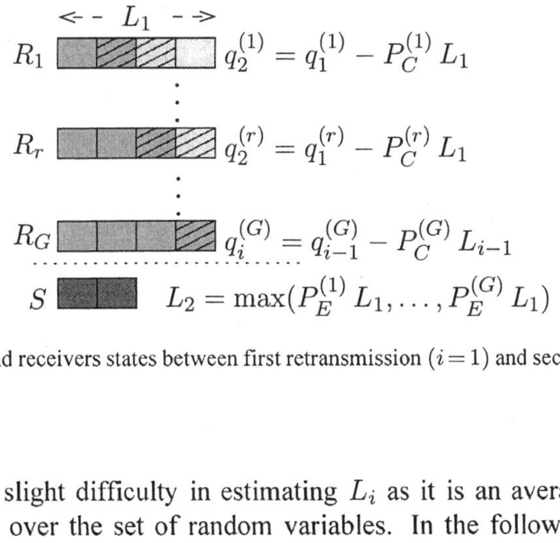

$\leftarrow - \; L_1 \; - \rightarrow$

R_1 $q_2^{(1)} = q_1^{(1)} - P_C^{(1)} L_1$

R_r $q_2^{(r)} = q_1^{(r)} - P_C^{(r)} L_1$

R_G $q_i^{(G)} = q_{i-1}^{(G)} - P_C^{(G)} L_{i-1}$

S $L_2 = \max(P_E^{(1)} L_1, \ldots, P_E^{(G)} L_1)$

Figure 12-5. The sender and receivers states between first retransmission ($i=1$) and second retransmission ($i=2$)

However, there is a slight difficulty in estimating L_i as it is an average of the maximum value over the set of random variables. In the following paragraph we derive the probability density function and the average of $\max(X_1, \ldots, X_G)$, where $X_1, \ldots, X_G$ are independent and non-identically distributed random variables.

Determining $L_{i+1} = E[X_{\max}, i]$. Let the sender transmit L_i packets to the set of G receivers. Each receiver may have 0 to L_i packets erased. Let G random variables, $X_1, \ldots, X_G$, denote the number of erased packets for the receivers in the set. Then we can define $X_{\max}$ as follows:

DEFINITION 12.1 *$X_{\max} = k$ is an event that at least one receiver in the set loses k packets and no receiver from the set loses more than k packets, in round i, where $0 \le k \le L_i$.*

The probability that a receiver r loses exactly j packets if L_i was transmitted, $P_{E,i,j}^{(r)}$, is given by Eq. (12.19), and its cumulative density function is simply

$$F_r(j,i) = \Pr[X_r \leq j, i]\,. \tag{12.21}$$

Starting from Def. 12.1, it can be shown that for every k and i,

$$\Pr[X_{\max} = k, i] = \begin{cases} \prod_{r=1}^{G} F_r(k,i) - \prod_{r=1}^{G} F_r(k-1,i), & 1 \leq k \leq L_i \\ \prod_{r=1}^{G} F_r(k,i), & k = 0 \\ 0, & \text{otherwise.} \end{cases} \tag{12.22}$$

Eq. (12.22) defines the probability density function for event described by Def. 12.1. Finally, the mean value of $X_{\max}$ is

$$E[X_{\max}, i] = \sum_{k=0}^{L_i} k \Pr[X_{\max} = k, i] = L_{i+1}. \tag{12.23}$$

4.3 Calculation of G_i for multicast with ideal hybrid ARQ

In hybrid ARQ, as well as in ARQ scheme, when a receiver r confirms the file reception, it leaves the multicast group. The probability that a receiver r will leave the multicast group after round i is $P_{E,i,0}^{(r)}$, as defined in (12.19). Then, the probability that the receiver r would remain in the group in $i+1$ is $1 - P_{E,i,0}^{(r)}$.
Therefore, to calculate the average size of the multicast group in round i for hybrid ARQ case, we can generally apply (12.15) – (12.18), described in section 3.2. The difference is that here we replace $P_{EL,i-1}^{(r)}$ with $P_{E,i-1,0}^{(r)}$ and $P_{CL,i-1}^{(r)}$ with $1 - P_{E,i-1,0}^{(r)}$, where $P_{EL,i-1}^{(r)}$ is a realisation of pdf given in (12.19) for $i = i-1$ and $j = 0$.

5. SPECIFIC EXAMPLES AND PERFORMANCE COMPARISON

In this section we look at some examples of a network with i.i.d. channels and compare the analytical and simulation results. Simulations were carried using OPNET [1]. The main performance measure is a **sender throughput**. The sender throughput (12.24) is defined as a ratio of the file size and the

[1] http://www.opnet.com

total amount of data transmitted by a sender until all receivers in the network recover the file.

$$\text{Throughput} = \frac{L \cdot l}{\sum_{i=0}^{\infty}(L_i\,(l+h) + ap_i + o)} \tag{12.24}$$

where,

l	is the data packet payload size and $L \cdot l$ is the file size;
h	is the size of the data packet header;
ap_i	$= ap(G_i)$ is the size of the Address_PDU packet in round i, and it is proportional to G_i;
o	represents additional overheads including timeouts, acknowledgments, and guard-times, normalised with transmission data rate.

5.1 Examples for i.i.d. channels

If all channels have Bernoulli distribution of loss with average bit error rate ber_r for channel r, the probability of a data packet being erased in the channel r, $P_E^{(r)}$, is

$$P_E^{(r)} = 1 - (1 - ber_r)^{l+h} = 1 - P_C^{(r)}, \tag{12.25}$$

where $P_C^{(r)}$ is a probability that the packet is correctly received. If all channels are independent and identically distributed, then $\forall r = 1 \ldots G$ and $\forall i > 0$, $ber = ber_r$,

$$P_E^{(r)} = P_{ke} = 1 - P_{kc}, \tag{12.26}$$

and,

$$q_i^{(r)} = q_i.$$

Multicast with ARQ. The average L_i in the case of the multicast with ARQ only, is

$$L_i = L_{i-1}\,(1 - (1 - P_{ke}^i \frac{L}{L_{i-1}})^G)\,.$$

Furthermore, (12.13) and (12.14) become:

$$P_{CL,i-1} = P_{kc}^{q_{i-1}} \quad \text{and} \quad P_{EL,i-1} = 1 - P_{kc}^{q_{i-1}}.$$

Substituting them into (12.15) – (12.17), the probability density function for the remaining number of active receivers in the group in round i simplifies to:

$$
\begin{aligned}
p_0^{(i)} &= P_{kc}^{q_{i-1}G}\,, \\
p_1^{(i)} &= G\,(1 - P_{kc}^{q_{i-1}})\,P_{kc}^{q_{i-1}(G-1)}\,, \\
&\vdots \\
p_g^{(i)} &= \tbinom{G}{g}\,(1 - P_{kc}^{q_{i-1}})^g\,P_{kc}^{q_{i-1}(G-g)}\,, \\
&\vdots \\
p_G^{(i)} &= (1 - P_{kc}^{q_{i-1}})^G.
\end{aligned}
$$

From the above it is obvious that the number of unsuccessful receivers in round i has a binomial distribution with mean

$$
\begin{aligned}
G_i &= GP_{EL,i-1} \\
&= G(1 - P_{kc}^{q_{i-1}})\,,\ \forall i > 0\,.
\end{aligned} \tag{12.27}
$$

Fig. 12-6 shows the sender throughput obtained using simulations and the throughput obtained applying the above equations. The throughput is presented as a function of data packet payload, l, for various values of bit error rates ber. For every value of l and ber the size of the transmitted file is the same, therefore L varies with l. The group size is 30 receivers, $h = 16$ bytes, $ap_i = 24 + 8G_i$ bytes, $o = 90$ bytes on average. It can be noticed that theoretical and simulation plots completely overlap.

Multicast with ideal hybrid ARQ. If all channels are i.i.d., the equations in Section 4 may simplify considerably. Here, (12.22) becomes

$$
\Pr[X_{\max}=k, i] = \begin{cases} (F_r(k,i))^G - (F_r(k-1,i))^G, & 1 \le k \le L_i \\ (F_r(k,i))^G, & k = 0 \\ 0. & \text{otherwise,} \end{cases}
$$

The average number of parity packets transmitted in round $i+1$, which is the average value for the above distribution, simplifies to:

$$
L_{i+1} = E[X_{\max}, i] = L_i - \sum_{k=0}^{L_i-1} (F_r(k,i))^G.
$$

Here, $F_r(k,i)$ is calculated from (12.21) and (12.19) by substituting $P_E^{(r)}$ and $P_C^{(r)}$ with P_{ke} and $1 - P_{ke}$ given in (12.25)–(12.26). Furthermore, the average number of packets that each receiver expects is

$$
q_{i+1} = q_i - P_{kc}\,L_i.
$$

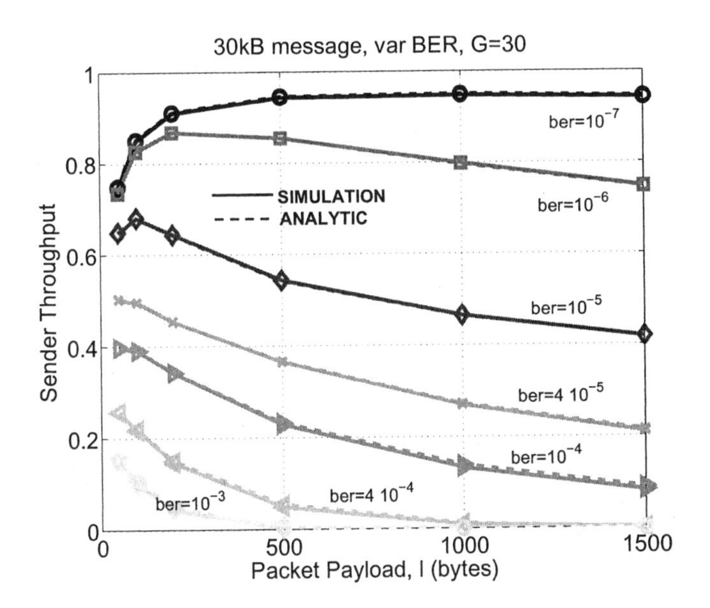

Figure 12-6. Sender throughput: simulation vs analytical results for multicast ARQ

Similarly to (12.27), the average number of receivers in the next round of multicast with ideal hybrid ARQ error recovery here becomes

$$G_{i+1} = G(1 - P^{(r)}_{E,i,0}).$$

In this case, $P_{E,i,0}$, defined in (12.19), simplifies to

$$P_{E,i,0} = \sum_{k=0}^{L_i - q_i} \binom{L_i}{k} (1 - P_{kc})^k \, (P_{kc})^{L_i - k}.$$

Fig. 12-7 compares the sender throughput obtained using simulations and the throughput obtained applying the above equations. The throughput is presented as a function of data packet payload, l, for various values of bit error rates ber. All file and network parameters are identical to the scenario with ARQ. The simulations and analytical plots follow each other very closely. One of the reasons for the small discrepancies may come from the rounding off errors when estimating the values of L_i and q_i in each retransmission round. This was not happening in the analytical model for ARQ. The gradual divergence of the theoretical and simulation curves for large packet sizes may be caused by the fact that L is inversely proportional to the packet payload size, l. Small value of L may lead to the coarser estimations of the average L_i. The similar may have happened with the simulation results when the packet loss rate is very high.

Multicast with ARQ vs multicast with hybrid ARQ Finally, Fig. 12-8 shows a comparison between multicast using ARQ error recovery and ideal hybrid ARQ error recovery. The hybrid ARQ shows exceptional dominance over ARQ especially for large packet error rates. For example, let us observe curves for $ber = 4 \cdot 10^{-5}$ while varying the payload size, l. Corresponding packet error rates would be approximately in the range 10^{-2}– 0.4, according to (12.25) and (12.26). It can be noticed that the throughput for multicast with ideal hybrid ARQ exhibits close to 100% improvement when compared to the multicast with ARQ only. For instance, at $ber = 4 \cdot 10^{-5}$ and $l = 500$ bytes, ideal FEC/ARQ (hybrid ARQ) throughput is about 0.72, while ARQ only case gives throughput of about 0.38.

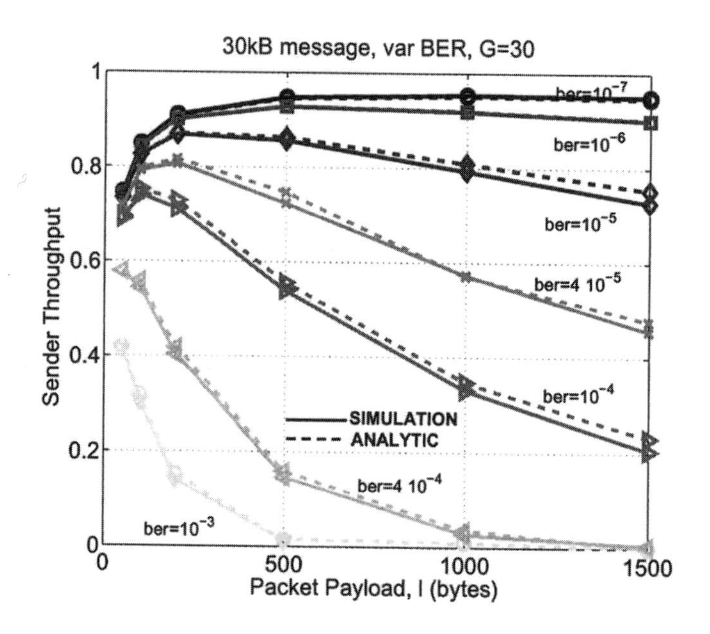

Figure 12-7. Sender throughput: simulation vs analytical results for multicast ARQ

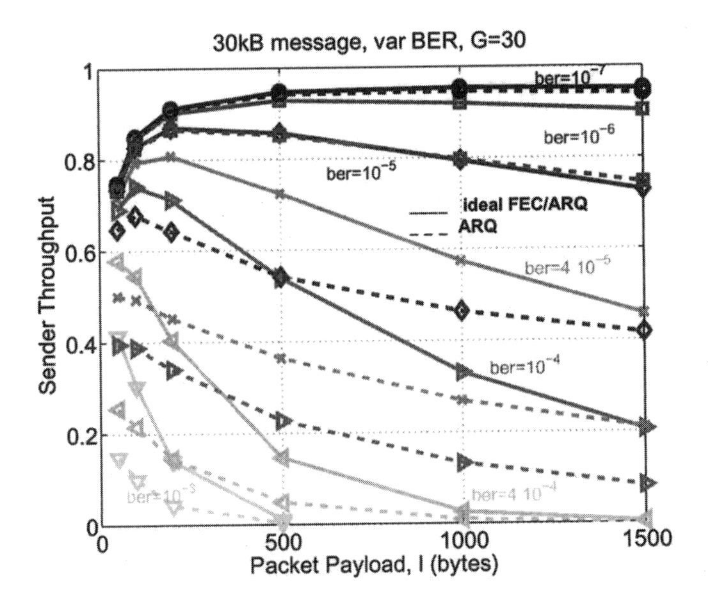

Figure 12-8. Comparison of the sender throughput for multicast with ARQ and multicast with ideal hybrid ARQ (a.k.a. ideal FEC/ARQ) error recovery mechanisms

6. CONCLUSION

We have presented a theoretical method to calculate the throughput for message-based reliable multicast protocols with 1) pure ARQ error recovery scheme, and 2) ideal hybrid ARQ (ARQ with an ideal rateless FEC code) error recovery scheme. For an arbitrary retransmission round, i, we derived expressions for the average number of requested packets by a receiver, q_i, the average reduced message size, L_i, and the average reduced multicast group size, G_i, which all may be used and extended to other multicast and unicast protocols. The calculations have shown good agreement with simulation results for i.i.d. channels. This work may be expanded to cover scenarios with diverse channel characteristics and other error recovery mechanisms.

7. REFERENCES

1. C. Diot et al. Deployment Issues for the IP Multicast Service and Architecture. *IEEE Network*, 14(1):78–88, Jan/Feb. 2000.

2. S. Pingali. *Protocol and Real-Time Scheduling Issues for Multimedia Applications*. PhD thesis, University of Massachusetts Amherst, Sep. 1994.

3. D. Towsley, J. Kurose, and S. Pingali. A Comparison of Sender-Initiated and Receiver-Initiated Reliable Multicast Protocols. *IEEE JSAC*, pages 398 – 406, Apr. 1997.

4. G.S. Poo and A.M. Goscinski. A Comparison of sender-based and receiver-based reliable multicast protocols. *Computer Comms.*, (21):597–605, 1998.

5. B. N. Levine and J. J. Garcia-Luna-Aceves. A comparison of reliable multicast protocols. *ACM Multimedia Systems*, 6(5):334–348, 1998.

6. J. Nonnenmacher, E. Biersack, and D. Towsley. Parity-based loss recovery for reliable multicast transmission. In *Proc. of SIGCOMM'97*, pages 289–300, Cannes, France, Sep. 1997.

7. C. Riechmann. P_MUL: An Application Protocol for the Reliable Data Transfer over Multicast Subnetworks and under EMCON Restrictions. IETF Internet draft, Apr. 1999. Work in progress.

8. M. Luby. LT codes. In *Proc. of The* 43th *Annual IEEE Symposium on Foundations of Computer Science*, pages 271–280, Nov. 2002.

9. P. Maymounkov. Online Codes. Technical Report TR2002-833, New York University, Nov. 2002.

10. J.W Byers, M. Luby, and M. Mitzenmacher. A digital fountain approach to asynchronous reliable multicast. *IEEE JSAC*, 20(8):1528 – 1540, Oct. 2002.

Chapter 13

FAIR QUEUING IN ACTIVE AND PROGRAMMABLE NETWORKS

Fariza Sabrina and Sanjay Jha
School of Computer Science and Engineering, The University of New South Wales, NSW 2052, Australia

Abstract At present the Internet is being transformed to a sophisticated system where network researchers are exploring new ways to dynamically program network switches, routers to accelerate network innovation. This trend introduces the concept of Active and programmable networks. The goal is to simplify the deployment of network services, leading to networks that explicitly support the process of service creation and deployment. To use such technology safely and efficiently, individual nodes must provide mechanisms to enforce resource limits to the contending flows. In active and programmable networks, the packet scheduling schemes should consider multiple resources such as CPU and memory in addition to bandwidth to achieve overall fairness. Maintaining fairness of one resource allocation does not ensure the fair allocation of other resources automatically. The dynamic nature of network load, and the unpredictability of processing times of active packets pose another significant challenge in providing fairness guarantees between contending flows. This chapter presents a fair resource allocation mechanism for a programmable/active node.

Keywords: Fair Resource Allocation, Scheduling, Programmable Networks, Active Networks.

1. INTRODUCTION

Active network is a framework where network nodes not only forward packets, but also perform customized computation on the packets flowing through them. The goal is to make the network a more general computation engine. It provides a programmable interface to the user where users dynamically inject services into the intermediate nodes. Intermediate nodes that provide these services are called active nodes and the network packets that invoke the services are called active packets. The behaviors of active packets are controlled by the users through the programmable interface or via the packets they send. In these network when the packets arrive at a node, it first executes the program associated with the packet and then routes the packet to the next node. Active network technology envisions deployment of virtual execution environment within network elements, such as switches and routers.

In an Active network one of the fundamental problems is the development of efficient resource distribution model. While there has been significant research in developing suitable active node architectures [2–4, 6], programming models for active software, and developing efficient security architectures [7, 8], the problem of efficient resource allocation in active nodes has not been sufficiently addressed. Active and programmable networks facilitate fast deployment of new services and protocols by allowing execution of packets in routers and switches. They also allow customized computation on packets flowing through them [1–3]. To use this technology safely and efficiently, individual nodes must understand the varying resource requirements for specific network traffic and must manage their resources efficiently. This implies that each node must understand the varying resource requirements for specific network traffic.

In Active/programmable networks, packets belonging to different traffic flows share multiple resources among them. Fairness is an intuitively desirable property in the allocation of available resources among multiple traffic flows. Fairness in resources (e.g., CPU and Bandwidth) scheduling becomes especially desirable with the increasing use of parallel systems in multiuser environments with the interconnection network shared by several users at the same time. In multiuser environments, the protection guaranteed by fair scheduling of packets can improve the Quality of Service which is strictly desired by customers of parallel systems. Strict fairness is also desirable since unfair treatment of some traffic flows in the network can easily lead to unnecessary bottlenecks.

In traditional networks, several packet scheduling algorithms exist that aim to isolate different network flows from ill-behaved flows. The basis for isolation is derived through fair allocation of network resources (consisting only of bandwidth for core networks) [9–11]. Service disciplines such as Fair Queuing (FQ) provide perfect fairness among contending network flows. However, the traditional notion of fair queuing that specifies a resource allocation constraint for a single resource does not directly extend to active and programmable networks, since allocation of resources in these networks involves more than one resource, such as CPU, bandwidth, and memory. Moreover, the allocation of these resources is interdependent, and maintaining fairness in one resource allocation does not entail fairness for other resources [7]. It has been identified that CPU requirement of active packets on different platforms cannot be determined a priori and this is a major obstacle in managing the CPU resource in active nodes [13, 7].

It is apparent that for large-scale deployment of active and programmable networks, researchers should address the issues of managing multiple resources within a node. Ramachandran et al. [7] developed a packet scheduling algorithm to achieve fairness in active nodes, which satisfies the resource allocation constraint by adjusting the share of CPU resource given to each flow based on the share of bandwidth given to the flow. This was done with a separate CPU and bandwidth schedulers (based on Deficit Round Robin, or DRR), where the bandwidth scheduler periodically provided feedback to the CPU scheduler about the bandwidth consumed by the competing flows. The feedback mechanism was computationally expensive and the accuracy of maintaining fairness depended on the frequency of the feedback.

Pappu et al. [14] presented a processor scheduling algorithm called Estimation-based Fair Queuing (EFQ) that estimated the execution times of various applications on packet of given lengths and then scheduled the processing resources based on the estimation. They claimed that this algorithm provided better delay guarantees than processor scheduling algorithms that did not take packet execution times into consideration.

Galtier et al. [13] proposed a scheme to predict the CPU requirements to execute a specific code on a variety of platforms. In this work, a code can be executed on a specific platform off-line and the CPU requirement on this platform can be used to predict the CPU requirement on other platforms. They calculated the CPU requirement based on the relative calibration performance of the local and reference nodes. In reality, the predicted CPU requirement on a platform can differ significantly from the actual requirement. Therefore, it would be very difficult to implement the scheme on a busy node where allocation of proper share to competing flows was crucial.

In this chapter a fair resource allocation scheduler for active and programmable networks is presented. The presented scheduler can fairly allocate CPU and bandwidth resources to all the competing flows within a programmable or active node. This chapter is organized as follows: In section 2, we describe the desirable properties of a scheduling algorithm. In section 3, we present a node resource management architecture. Section 4 describes the CBCS scheduler, and the scheduling algorithm. In section 5 we provide simulation results, and the chapter concludes in section 6.

2. DESIRABLE PROPERTIES OF A SCHEDULING ALGORITHM

Future Internet is expected to support a variety of services. The network must ensure that each flow of traffic receives its fair share of resources and is able to provide performance guarantees. This requires an efficient resource management mechanism to apportion, allocate and manage limited resources efficiently and fairly among all the competing users. Some of the most important and desirable properties of a scheduling discipline are described below.

- *Fairness:* To ensure that the performance achieved by a flow is not affected when a misbehaving flow tries to consume more bandwidth and/or CPU resources than its fair share, the available resources (e.g., CPU and link bandwidth) must be fairly distributed among the flows sharing the link. Scheduling algorithms should be able to allocate resources by a certain scheme based on a particular notion of fairness to achieve performance guarantee required by applications.

- *Delay:* Many applications such as multimedia and teleconferencing applications are sentitive to delay and delay jitter. So delay is an important property that should be kept in mind while designing a scheduling algorithm. Resource scheduling scheme should be efficient enough to provide low end-to-end delay for the flows.

- *Complexity:* The simplicity and time complexity properties always collide with the fairness and delay bound properties. Schedulers with short-term fairness and strict delay bound generally have high time complexity and are hard to be implemented. O(1) time complexity schemes are easy to be implemented, but they are generally fair to provide short-term fairness and low delay bound.

Generally, packet scheduler should have the following properties:

1 Low complexity ;

2 Treats different flows fairly;

3 Provides low worst case delay and delay jitter;

3. MANAGING ACTIVE NODE RESOURCES

[12] presents a framework for resource management for active networks (as shown in Fig. 13-1). The architecture supports code-carrying active packets and

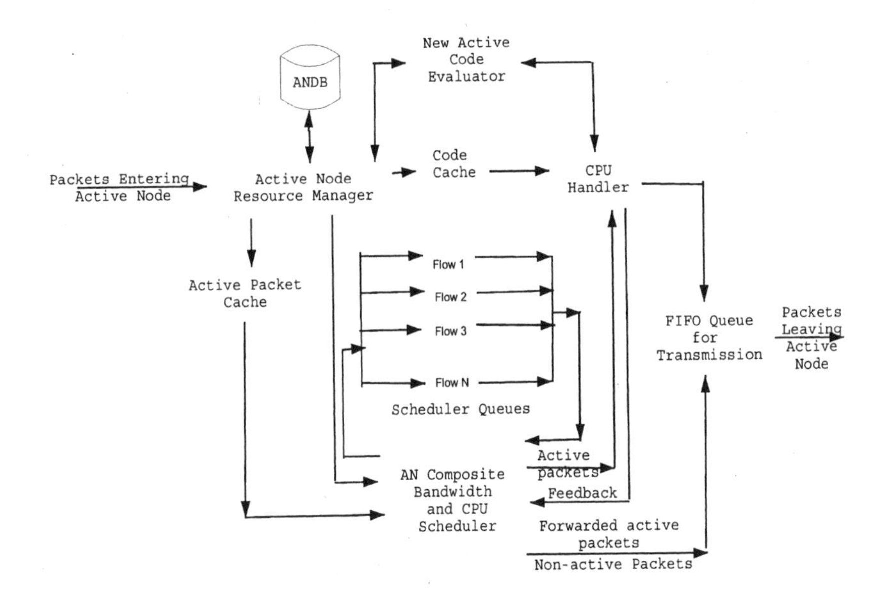

Figure 13-1. Active Node Resource Management Scheme

active packets that only carry references to certain codes or programs (assuming they were pre-installed on the node), and also non-active packets. An active packet contains active code that is identified by a GUID (as in the PANTS [4] and ANTS [5] packet formats). The active packet header contains fields that are used to specify the referenced code, monitor the processing status, and identify the type of service required, and so on. This section presents a resource management scheme based on architecture in [12] for best effort flows without any reservations.

It is considered that in most cases a best-effort flow would reference a limited number of different programs or code that the source may have already evaluated in the node prior to the flow's commencing. It may be noted that considering a limited number of referenced codes within a flow maps well to most active and programmable networking applications (e.g., video encoding, fusion, online brokering, data filtering) that require executing the same specific code on many packets within a flow.

When a packet enters an active node, the Active Node Resource Manager (ANRM) reads the packet header to determine the service requirement of the packet. For an active packet that must be evaluated, ANRM uses the Active

Code Evaluator (ACE) to evaluate the CPU requirements of the packet against the referenced code. If the packet does not carry the referenced code and the code is not available in the code cache, then ANRM sends a request to the packet stream's source node to send the referenced code and puts the packet in a waiting queue in the active packet cache.

It may be noted that processing of packets on an active or programmable node can also affect the size of packets after processing is completed. To take this packet size change into account for bandwidth consumption, an Expansion Factor is calculated, that is the packet size after processing divided by the packet size before processing. After evaluation of the packet is done, ANRM updates the Active Node Database (ANDB) with the CPU requirement (against evaluated packet size) and the Expansion Factor for the referenced code.

For an active packet that must be processed in the node, the ANRM searches the ANDB for a record corresponding to the referenced code. If a record is found then it determines the CPU requirement based on the recorded information in the ANDB and passes the packet to the scheduler object, specifying the CPU requirement and the Expansion Factor. The CPU requirement of the packet may differ from the CPU requirement of the evaluated packet as recorded in the ANDB. Pappu et al. [14] have shown experimentally that for header processing applications (e.g., IP forwarding) the processing time does not depend on the packet size, while for payload processing applications (e.g., compression) the processing time varies linearly with the packet size. ANRM takes the current packet size and the evaluated packet size as recorded in ANDB into account for determining the CPU requirement if the referenced code were classified as a payload processing application in the ANDB record.

If a record was not found in ANDB for the referenced code, the ANRM checks whether the referenced code is available (either carried by the packet or in the code cache). If it is available, ANRM uses the ACE object to evaluate the packet, updates the ANDB and then passes the packet to the scheduler object, specifying the CPU requirement and Expansion Factor. If the referenced code is not available then the ANRM puts the packet in a waiting queue in the active packet cache and sends a request to the source node to send the referenced code. Note that this request may not traverse up to the source node; it may be intercepted by an active node on the way and the code found in that node. When the requested referenced code arrives in the node, ANRM evaluates the CPU requirement and updates the ANDB, and at this point all packets from the waiting queue corresponding to the referenced code are handed to the scheduler object. The non-active packets and the active packets that do not require processing in this node are handed to the scheduler object specifying zero CPU processing time and 1.0 as the Expansion Factor.

4. *CBCS* - A COMPOSITE BANDWIDTH AND CPU SCHEDULER BASED ON DRR PRINCIPLE

CBCS is especially designed to serve best-effort flows and it guarantees fair shares of both CPU and bandwidth between all the competing flows. The scheduler limits the resource allocation to an individual flow by continuously monitoring the individual queue lengths in terms of the total CPU requirement for all packets in each queue. If the individual queue length becomes greater than a pre-defined threshold (which is configurable for each flow) then the scheduler stops accepting packets from the corresponding flow. ANRM then puts these packets into the Active Packet Cache (which includes a separate FIFO queue for each flow) and they will be handed to the scheduler when the queue length gets smaller than another predefined threshold. The details of the algorithm are discussed below.

The scheduler object enqueues packets in the corresponding queues and dequeues packets using its fair-scheduling algorithm $CBCS$, which guarantees fair shares of both CPU and bandwidth between all the competing flows. After dequeuing a packet, the scheduler hands the packet to the CPU handler to run the code in the CPU. The packet is sent to its next destination after processing. The scheduler also continuously monitors the individual queue lengths in terms of the total CPU requirement for all packets in each queue. If the individual queue length becomes greater than a pre-defined threshold (which is configurable for each flow) then the scheduler stops accepting packets from the corresponding flow. ANRM then puts these packets into the Active Packet Cache (which includes a separate FIFO queue for each flow) and they will be handed to the scheduler when the queue length gets smaller than another predefined threshold. The details of the algorithm are discussed below.

4.1 CBCS ALGORITHM

CBCS enforces the fairness using DRR (Deficit Round Robin) principle. It succeeds in eliminating the unfairness of pure packet-based round-robin by keeping a deficit counter to measure the past unfairness. The psuedo-code of the scheduling algorithm is shown in Fig. 13-2. The algorithm uses the following parameters and equations:

- $Quantum$ = A variable that represent a time slice that is used to serve packets from each flow queue (which includes both CPU processing time and network transmission time) (msec)
- $DeficitCounter[i]$ = A State variable that represents a time slice for which flow i queue deserves to be served within a specific round of scheduling (msec).
- BW = Bandwidth of the transmission link in Mbps.

- P^{Size} = Size of a specific packet within a competing flow in Bits.
- P^{CPUReq} = CPU time requirement to process a specific packet in msec.
- CPU_t^{UL} = Upper limit of the total CPU Queue in terms of CPU processing time requirement for all the packets in all the flows.
- CPU_t^{LL} = Lower limit of the total CPU Queue in terms of CPU processing time requirement for all the packets in all the flows.
- γ = Expansion factor of the packet.
- P^{ts} = Time slice required for processing and transmitting the packet (msec).
- $P^{ts}_{(Max)}$ = The maximum allowable time slice that a packet can have to cover the both CPU processing and network transmission. It is configurable, and should depend on the capability of an active node.

Therefore,

$$P^{ts} = \frac{1000 * \gamma * P^{Size}}{BW} + P^{CPUReq} \tag{13.1}$$

When the scheduler receives a packet, it looks at the header of the packet to determine the flow-id for the packet, notes the CPU requirement and Expansion Factor, and then stores the packet in the corresponding flow queue. However, the scheduler continues to monitor the queue length for all individual flows in terms of CPU time requirement and stops accepting packets from a flow if its queue length becomes greater than $CPU^{UL}_{t[i]}$. In this case the scheduler continues to refuse new packets from flow i until its queue length becomes smaller than $CPU^{LL}_{t[i]}$.

When the scheduler starts, the *Quantum* is set to $P^{ts}_{(Max)}$ and the *DeficitCounter* for all flows is set to zero. The scheduler continues to serve all non-empty queues within each round of processing. When it starts to serve a queue within a round, the *DeficitCounter* is set to *Quantum* plus the *Deficit Counter* of the previous round. The scheduler then dequeues a packet from the head of the queue and calculates the P_{ts} of the packet according to the Eq.(13.1). It sets the *Deficit Counter* to (*DeficitCounter* - P^{ts}) and hands the packet to the CPU Handler object for execution. The packet is sent to its next destination after processing. The scheduler stops serving a queue once the queue is empty or the deficit counter becomes zero or negative. It may be noted that the *DeficitCounter* for a non-active flow (i.e., a flow having no packets in the queue) is reset to zero.

Initialization: (Invoked when the scheduler is initialized)

```
{
Quantum= P^ts_Max;
for (i=0; i<n; i=i+1)
      DeficitCounter[i]=0;    /* For all the flows DC is set to 0 */
}
```

Enqueue: (Invoked when a packet arrives)

```
{
find the flow id;
calculate the CPU bandwidth requirements in terms time;
put the packet in respective queue;
}
```

Dequeue: {

```
while (TRUE) {
    for each flow(i) {
      if (the queue is not empty)
        {
          DeficitCounter[i] = DeficitCounter[i] + Quantum;
            while(DeficitCounter[i] > 0 and the Queue is not empty)
              {
                De-queue a packet;
                Read the packet header and calculate the Pts ;
                DeficitCounter[i] = DeficitCounter[i] - Pts ;
                Hand the packet to CPU Handler;
              }
                if (the queue is empty)
                    {
                      //Don't Accumulate DeficitCounter for a non Active flow
                      DeficitCounter[i] = 0;
                    }
        }
        else
        {
            //Don't Accumulate DeficitCounter for a non Active flow
            DeficitCounter[i] = 0;
        }
                    }
                  }
                }
```

Figure 13-2. Psuedo-code for $CBCS$

4.2 DEFINITION OF FAIRNESS

The total resource consumption by a best-effort flow in an active node is used as the basis for measuring fairness. The following notations are introduced. Let,

$CPU_i^t(t_2 - t_1)$ = The CPU time consumed by the packets in flow$i$ within a time period of $(t_2 - t_1)$.

$BW_i^t(t_2 - t_1)$= The transmission time consumed by the packet in flow$i$ within a time period of $(t_2 - t_1)$.

$CPU_{tot}^t(t_2 - t)$ = The total CPU time consumed by packets in all flows within the time period $(t2 - t1)$.

$BW_{tot}^t(t_2 - t_1)$= The total transmission time consumed by packets in all flows within the time period $(t_2 - t_1)$.

N = Number of competing flows within the time period $(t2 - t1)$.

Definition 1: A flow is backlogged during the time interval $(t2 - t1)$ if the queue for the flow is never empty during the interval.

Definition 2: Using the notation above, for any period of time interval $(t_2 - t_1)$ the total resource allocated to flow i is given by:

$$CPU_i^t(t_2 - t_1) + BW_i^t(t_2 - t_1)$$

Definition 3: Our fairness measure specifies that for any time period of (t2 - t1) the total resource allocation (i.e., CPU plus bandwidth) for each individual backlogged flow is a constant:

$$CPU_1^t(t_2 - t_1) + BW_1^t(t_2 - t_1) = CPU_2^t(t_2 - t_1) + BW_2^t(t_2 - t_1) = \cdots = CPU_n^t(t_2 - t_1) + BW_n^t(t_2 - t_1) = C \tag{13.2}$$

where, $C \leq \frac{2(t_2 - t_1)}{N}$

Definition 4: An active node achieves perfect fairness if total resource allocation for a time period of $(t2 - t\ 1)$ for all the backlogged flows is the same, ideally $\frac{1}{N} + \frac{1}{N} = \frac{2}{N}$.

Definition 5: The deviation from the ideal allocation for any flow i is defined as:

$$Dev_i^{(t2-t1)} = \frac{2}{N} - \left(\frac{CPU_i^t(t2-t1)}{CPU_{tot}^t(t2-t1)} + \frac{BW_i^t(t2-t1)}{BW_{tot}^t(t2-t1)} \right)$$

Definition 6: The error in maintaining fairness in resource allocation for any backlogged flow i for the time period $(t2 - t1)$ is defined as:

$$Error_i^{(t2-t1)} = \{Dev_i^{(t2-t1)}\}^2 \tag{13.3}$$

Definition 7: The error in maintaining fairness in resource allocation for any backlogged glow i for the time period $(t2 - t1)$ is measure as

$$Error_{avg}^{(t2-t1)} = \sum_{i=1}^{N} \frac{Error_i^{(t2-t1)}}{N} \tag{13.4}$$

4.3 PARALLELISM IN CBCS

CBCS is a two-staged scheduler. The scheduler de-queues a packet from an individual queue, calculates the P^{ts}, updates the deficit counter by deducting the P^{ts} and then gives the packet to the CPU Handler object. CPU Handler process the packet and then the packet is en-queued in a FIFO queue for Bandwidth transmission which is served by a separate thread. i.e. while a packet is being processed in the CPU, it does not block the transmission of other packets from the transmission queue.

Lets take an example. Say, scheduler is serving two packets from input queues. First packet has P^{ts} = 10 ms (CPU = 2 ms and Bandwidth = 8 ms), and the second packet also has P^{ts} = 10 ms (CPU = 8 ms and Bandwidth = 2 ms). The scheduler picks the first packet, and hands it to CPU handler. CPU handler process the packet for 2 ms and then it puts the packet in the transmission queue. After putting the packet in the transmission queue, scheduler picks the second packet and hands it to CPU handler for Processing. So after 2 ms, the scenario is: The second packet is being processed in the CPU, and the first packet is being transmitted through the link. Since the bandwidth transmission time and CPU time are already considered within P^{ts} for de-queuing the packets from the individual input queues, a single FIFO queue is used for Bandwidth transmission for all the packets.

4.4 WORK COMPLEXITY

The work complexity could be determined from the time complexity of enqueuing and then dequeuing a packet for transmission. As shown in the psuedo-code, the enqueue operation consist of determining the flow at which the packet arrives and adding the flow to the linked list if it is not already in the list. Both of the operation is O(1) operation. Now the time complexity of dequeuing a packet is considered. The dequeue procedure include determining the next flow to be served, calculating the deficit counter, removing the flow from the active list. All of these can be done in constant time, so we can say that dequeue operation is of time complexity O(1). So the work Complexity for $CBCS$ scheduler is O(1).

5. SIMULATION RESULTS

The effectiveness of the algorithm in achieving overall fairness is analysed through simulations. We have modified the NS2 network simulator [15] to implement the described active node components to simulate an active node-based network. The simulation was performed on a PC having a 1.5 GHZ Pentium 4 processor and 512 MB memory and running under the Linux operating system (RedHat 7.2). The experimental results achieved using CBCS are compared with the results using DRR for scheduling CPU and bandwidth independently. Note that the processing overhead and memory requirement for the proposed system (e.g., to classify and manage individual flows) are minimal. The state information of each individual active flow is kept in the memory (in arrays of integer and double variables).

5.1 DEFAULT SIMULATION SETTINGS

We used 10 hosts to generate network for an active node, where each host generated CBR traffic corresponding to a flow. Host 1 generated active packets containing MPEG2 video data and referencing MPEG2 encoder code. The other hosts generated active packets with different packet sizes and referencing different code GUIDs. MPEG2 code was implemented in C++ within the NS2 environment and the evaluator object used the code to evaluate the CPU requirement and expansion factor of MPEG2 data packet the first time one arrived at the active node. For the other hosts, the packets generated from a given host carried the same GUID, and the simulated evaluator object randomly evaluated the CPU requirements for the code (when the first packet arrived in the active node) between 1 and 10 milliseconds and the expansion factor was set to 1. The output link capacity was set to 5 Mbps. The simulation settings of the individual flows are given below.

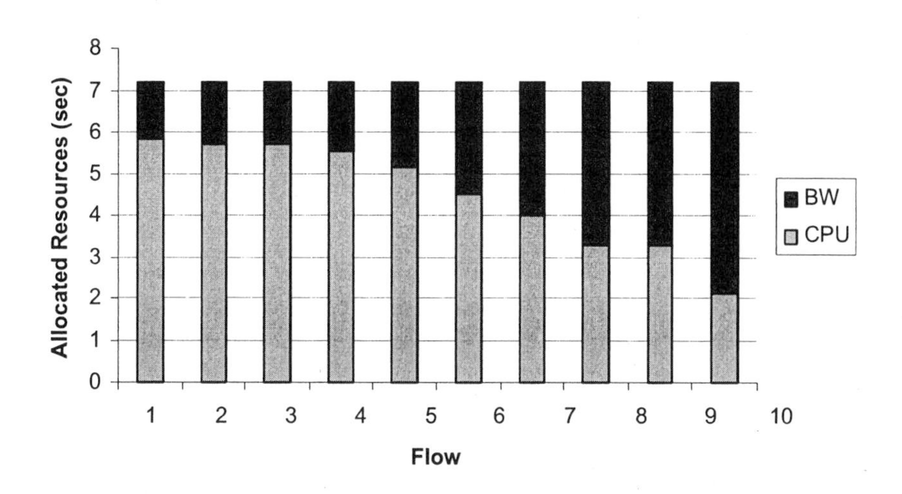

Figure 13-3. Total Resource Allocation at 45th sec using CBCS

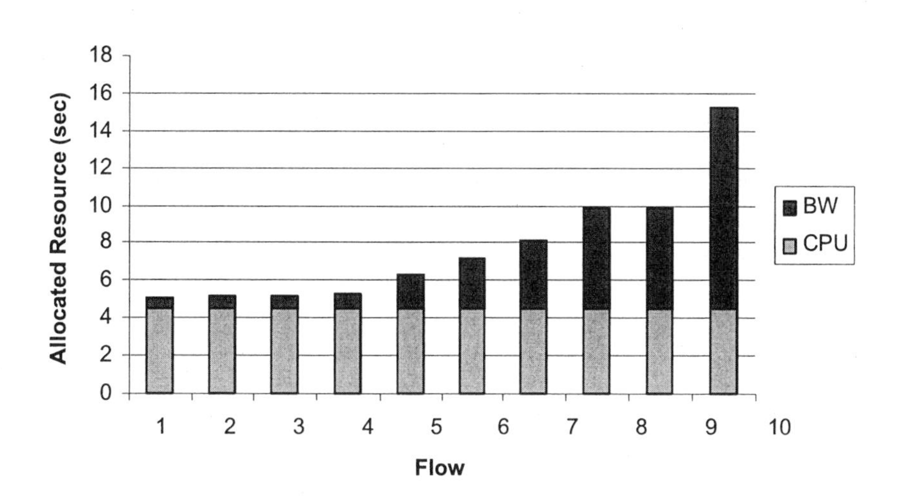

Figure 13-4. Total Resource Allocation at 45th sec using DRR

The simulations were run for 50 seconds and measurements were taken at 5-second intervals. Packet generation rates for all the flows were adjusted such that the topology exhibited congestion or backlog for each flow at the active node for the entire simulation time of 50 seconds. Each simulated active packet header contained some additional fields (or parameters) to signify the packet as an active packet and to facilitate the handling of the packet by the active node. In reality, the communication between active nodes could be done by exchanging messages, similar to the PANTS [4] architecture. However, for our simulation in NS2 simulator we added some parameters in the common header (i.e., hdr_cmn structure) of the NS2 packet.

5.2 FAIRNESS MEASUREMENT

As mentioned earlier, the objective of our composite scheduling algorithm is to allocate CPU and bandwidth resources proportionately and fairly according to Definition 3. CPU allocation and bandwidth allocation were measured every 5 seconds. We found that while using CBCS, the total allocation of resources for each flow remained more or less constant at any time. As sample results, the measurements at the 45th second are shown in Fig. 13-3.

Fig. 13-3 demonstrates that CBCS maintained perfect fairness by allocating more CPU to flows 1-4 and more bandwidth to flows 7-10 while keeping the total allocation almost constant (which is $\frac{(45*2)}{10}$ or 9 seconds). For instance, the total allocation for individual flows varied between 8.97 and 8.99 seconds. The measured CPU utilization was 100% and bandwidth utilization was 99.97%

Fig. 13-4 show the resource allocation results using DRR separately for CPU scheduling and bandwidth scheduling. It demonstrates that maintaining fairness in CPU scheduling does not entail fairness in bandwidth scheduling. Though the CPU allocation between active flows remains fair, the bandwidth allocation varies.

We also measured the error (i.e., deviation from the ideal situation) as defined in Eq. (13.4) at 5-second intervals for all the flows. Average errors are shown in Fig. 13-5. The average error for each flow was recorded as 0.007 while using DRR. Thus, on average each flow deviates from the fair resource allocation constraints of 0.2 by 0.0837 or by 41.83%. When we use CBCS, the average error reaches almost zero. Thus, using CBCS the achieved results did not deviate noticeably from the expected ideal situation.

5.3 DELAYS AND THROUGHPUT

We measured delays and throughput under different simulation settings, where all the flows required 97.8% of the CPU resources and 98.9% of the bandwidth resources (i.e., resource utilizations were just below 100%) so that

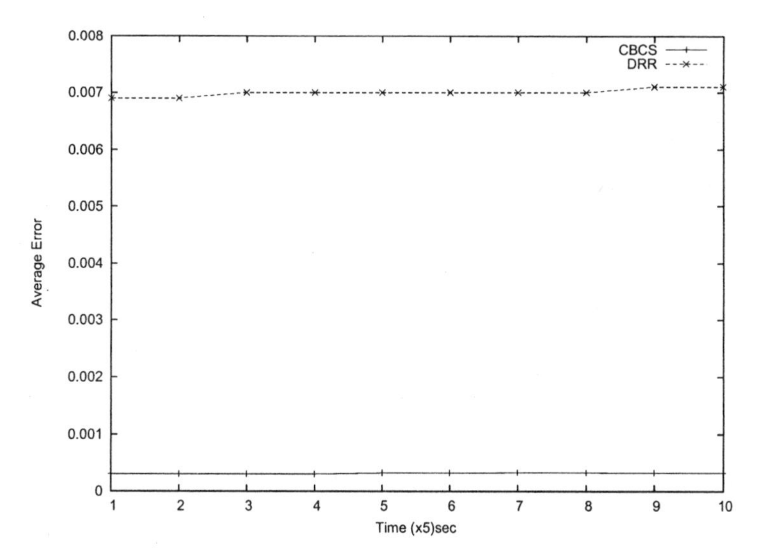

Figure 13-5. Average Error

the measured delays are because of scheduling and not because of queuing backlog. Though the measured throughput of each flow remained the same under DRR and CBCS, the measured delays showed that CBCS could provide somewhat better delay guarantees. For example, the delay plots for the MPEG data traffic are shown in Fig. 13-6 and 13-7. Using DRR the delays range from 0.142 to 0.316 Sec whereas using CBCS they vary from 0.142 to 0.250 Sec.

We also measured the throughput when all the flows were highly congested or backlogged, and the results are shown in Table 13-1. It shows that under highly congested scenario CBCS can provide significant improvement on throughput and proper utilization of resources compared to using DRR separately for CPU and bandwidth scheduling. With DRR, a significant number of packets from flows 8-10 were dropped after being processed, thus wasting the consumed CPU resources.

6. CONCLUSION

Resource management is a major challenge for active and programmable networks. An intuitively desirable and also a practically important property of a packet scheduler is the fairness it achieves in the allocation of resources. Fairness in scheduling is essential to protect flows from other misbehaving flows. Unlike traditional network, in active/programmable networks, fair allocation of bandwidth resource is not enough, it must also ensure fair CPU resource

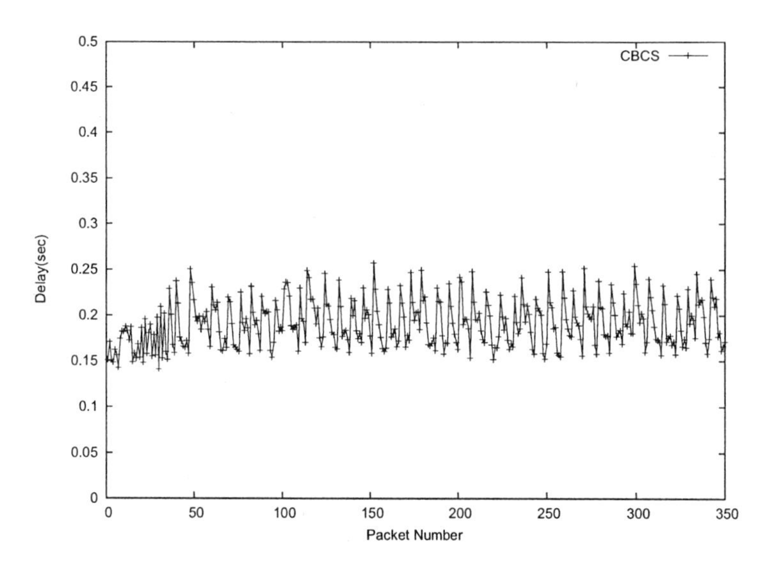

Figure 13-6. Delay for MPEG2 data traffic using CBCS

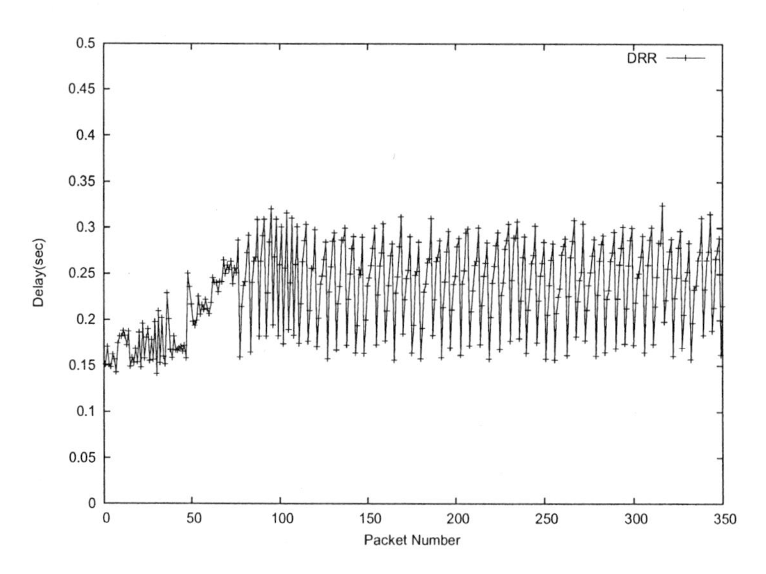

Figure 13-7. Delay for MPEG2 data traffic using DRR

allocation among contending flows. This chapter presented a composite fair scheduling algorithm that could schedule both CPU and bandwidth resources

Table 13-1. Fairness Measurements.

Flow Number	*packet entering Active Node*	*Packet processed at Active Node*		*Packet Received at Sink*	
		CBCS	DRR	CBCS	DRR
1	445	373	226	371	226
2	750	614	451	611	451
3	750	614	451	611	451
4	938	711	565	707	565
5	1050	841	751	838	751
6	1332	1031	1121	1028	1121
7	1623	1164	1495	1161	1495
8	2355	1335	2242	1332	1227
9	2352	1335	2243	1328	838
10	4820	1565	4491	1556	1378

adaptively and fairly among all the competing flows. We evaluated the performance of our combined scheduling algorithm for fair service among all the competing flows through simulation. The simulation results show that our composite scheduling algorithm offers better fairness, throughput, and delays than the traditional DRR algorithm if was used separately for CPU and bandwidth scheduling.

ACKNOWLEDGMENTS

This work is funded by Smart Internet Technology Cooperative Research Centre (SITCRC), Australia

REFERENCES

1. A. Campbell, H. G De Meer, M. E. Kounavis, "A survey of Programmable Networks", Center for Telecommunication Research, Columbia university, *ACM SIGGCOMM Comp. Commun. Rev.*, April 1999.
2. D. Tenenhouse, J. M.Smith, W. D.Sincoskie, D. J.Wetherall, and G. J.Minden, "A Survey of active network research", *IEEE communications Magaziné*, pp.80-86, Jan 1997.
3. D. Wetherall, U. Legedza, and J. Guttang, "Introducing new internet services:why and how," *IEEE Network Special Issue on Active and Programmable Networks*, 12(3):12-19, July 1998.
4. A. Fernando and B Kummerfeld, "Pants: Python active node transfer system,", *Tech Rep.*, University of Sydney, Australia, 1998
5. D.Wetherall, J.Guttang, and D.Tenenhousr, "Ants:A toolkit for building and dynamically deploying network protocols," 1st *IEEE OPENARCH 98*, pp 117-129, San Fransisco, CA, April 1998.

6. X. Qie, A. Bavier, L. Peterson, and Scott and Karlin, "Scheduling computations on a software-based router," in proc. *IEEE Joint International Conference on Measurement Modeling of Computer Systems* (*SIGMETRICS*), pp 13-24 Cambridge, MA, June 2001, IEEE.
7. V. Ramachandran, R. Pandey and S. H. Chan, "Fair Resource Allocation in Active Networks," in *Proceedings of the IEEE International Conference on Computer Communications and Networks* (*ICCCN*), pp. 468-475, Las Vegas, Nevada, Oct 16-18, 2000.
8. A.Demers, S. Keshav, and S, Shenker. "Analysis and simulation of a fair queueing algorithm," In *SIGCOMM '89*, pp 1-12, September 1989.
9. A.K. Parekh and R.G.Gallagher. "A generalized processor sharing approach in integrated services networks," In *INFOCOMM '93*, 1993.
10. I. Stoica, S. Shenker, and H. Zhang. "Core-stateless Fair queuing:A Scalable Architecture to Approximate Fair Bandwidth Allocations in High Speed Networks," In *Proc. of ACM SIGCOMM*, August 1998.
11. M. Shreedhar and G. Varghese. "Efficient Fair queueing using Deficit round robin," In *SIGCOMM '95*, August 1995.
12. F. Sabrina, S. Jha, "An Adaptive Resource Management Architecture for Active Network", *Kluwer Journal on Telecommunication Systems* 2003, pp 139-166, Kluwer Academic Publishers, 2003, 24(2-4)
13. V. Galtier, K. Mills, "Predicting and Controlling Resource Usage in a Heterogeneous Active Network," In proc. of *3rd Annual International Workshop on Active Middleware Services* (*AMS 2001*), pp 35-44, San Francisco, CA, USA, 6 August 2001
14. P. Pappu and T. Wolf, "Scheduling Processing Resources in Programmable Routers, Department of Computer Science, Washington University in St. Louis, MO, USA, WUCS-01-32, July 25, 2001,"
15. NS2 simulator, The LBLN Network Simulator, University of California, Berkley, USA.

INDEX